RECENT DEVELOPMENTS IN PHOTOVOLTAIC MATERIALS AND DEVICES

Edited by **Natarajan Prabaharan, Marc A. Rosen** and **Pietro Elia Campana**

Recent Developments in Photovoltaic Materials and Devices
http://dx.doi.org/10.5772/intechopen.74107
Edited by Natarajan Prabaharan, Marc A. Rosen and Pietro Elia Campana

Contributors

Himanshu Dehra, Ali Reza Reisi, Ashkan Alidousti, Djamel Eddine Tourqui, Achour Betka, Atallah Smaili, Tayeb Allaoui, Chukwuemeka Ikedi, Koichiro Yamauchi, Sang Hee Lee, Soo Hong Lee, Williams S. Ebhota, Tien-Chien Jen

First published in London, United Kingdom, 2019 by IntechOpen
IntechOpen is the global imprint of INTECHOPEN LIMITED, registered in England and Wales, registration number: 11086078, The Shard, 25th floor, 32 London Bridge Street
London, SE19SG – United Kingdom
Printed in Croatia

British Library Cataloguing-in-Publication Data
A catalogue record for this book is available from the British Library

Additional hard copies can be obtained from orders@intechopen.com

Recent Developments in Photovoltaic Materials and Devices, Edited by Natarajan Prabaharan, Marc A. Rosen and Pietro Elia Campana
p. cm.
Print ISBN 978-1-78985-403-9
Online ISBN 978-1-78985-404-6

For EU product safety concerns:
IN TECH d.o.o., Prolaz Marije Krucifikse Kozulić 3, 51000 Rijeka, Croatia,
info@intechopen.com or visit our website at intechopen.com.

Meet the editors

Dr. Natarajan Prabaharan received his BE degree in Electrical and Electronics Engineering from Anna University, Chennai, India, in 2012 and his ME degree in Power Electronics and Drives from Anna University, Chennai, India, in 2014. He obtained the University Merit Ranker Award in 2014 and received his PhD degree in Energy and Power Electronics from VIT University, Vellore, India, in 2017. He is currently as an assistant professor in the Department of Electrical and Electronics Engineering at SASTRA Deemed University, Thanjavur, Tamilnadu, India. He serves as an associate editor for *IET Renewable Power Generation, IEEE Access, Journal of Power Electronics,* and *International Journal of Renewable Energy Research*. He is a technical program committee member for various international conferences. Also, he is a reviewer for various reputed journals, including IEEE, IET, Elsevier, Springer, and Taylor and Francis. His research interest includes power electronics, new topologies for inverter and converters, grid integration of renewable energy sources and its controllers, photovoltaic systems, and microgrids.

Dr. Marc A. Rosen is a professor at the University of Ontario Institute of Technology in Oshawa, Canada, where he served as founding Dean of the Faculty of Engineering and Applied Science. Dr. Rosen has served as President of the Engineering Institute of Canada and President of the Canadian Society for Mechanical Engineering. He is a registered professional engineer in Ontario, and has served in many professional capacities, including editor-in-chief of several journals and a member of the board of directors of Oshawa Power and Utilities Corporation. With over 60 research grants and contracts and 700 technical publications, Dr. Rosen is an active teacher and researcher in sustainable energy, the environmental impact of energy and industrial systems, and energy technology (including renewable energy and efficiency improvement). Much of his research has been carried out for industry, and he has written numerous books. Dr. Rosen has worked for organizations such as Imatra Power Company in Finland, Argonne National Laboratory near Chicago, and the Institute for Hydrogen Systems near Toronto. Dr. Rosen has received numerous awards and honors, and he is a Fellow of the Engineering Institute of Canada, the Canadian Academy of Engineering, the Canadian Society for Mechanical Engineering, the American Society of Mechanical Engineers, the International Energy Foundation, and the Canadian Society for Senior Engineers.

Dr. Pietro Elia Campana received his Bachelor's and Master's degrees in Environmental Engineering both *cum laude* from the University of Perugia in Italy in 2009 and 2011, respectively. In 2014, he was a research assistant at the International Institute for Applied Systems Analysis in Austria. He received his PhD in Energy and Environmental Engineering from Mälardalen University in Sweden in 2015. Since 2015, he has been a research assistant professor at Mälardalen University. In 2017, he split his work time between Mälardalen University and KTH Royal Institute of Technology. Since 2016, he has been Assistant Editor of the Elsevier journal *Applied Energy*. His research fields include distributed energy systems, renewable energy, energy storage systems, multivector energy systems, energy system simulation/optimization, energy management strategies, energy in buildings, artificial intelligence, and water–food–energy nexus. He has been actively involved in 15 research projects and has more than 50 publications. He has also served as a reviewer for more than 40 international journals and five international conferences, and has been on the organizing and scientific committees for several international conferences.

Contents

Preface

The advancements in photovoltaic (PV) materials, devices, and ideas for developing efficient low-cost solar electric conversion are investigated in this book. PV cells from sheet silicon (monocrystalline, polycrystalline, or amorphous), gallium arsenide, metal chalcogenides and organometallics, and other thin-film materials, such as [Cd zn]S/CuInSe2, CdTe, and Zn3P2, have increased in their performance and efficiency during the past few years. Recently, mesoscopic solar cells have made an impact on commercial markets. Dye-sensitized solar cells and perovskite solar cells have seen exceptional growth in performance and commercialization. Many cells have exceeded or are rapidly approaching a solar-to-electric conversion efficiency of 10%. Advanced concentrator concepts, which are favorable for achieving greater than 30% conversion efficiencies using multijunction cells, are propitious. The flat-plate luminescent concentrator, while relatively low in efficiency, has many features that are alluring. Electrochemical PV cells that can generate electricity directly in a regenerative device or with the addition of a third electrode can store energy electrochemically and have exhibited an increase in durability.

Organic solar cells have become a subject undergoing intense study in industrial research. This is because solution-processable conjugated organic materials have the potential to enable simple fabrication of low-cost, mechanically flexible, and large area PV devices that allow relevant access to sustainable and clean energy from the sun. Significant effort has been devoted toward increasing the power conversion efficiency of such solar cells. A major breakthrough was achieved by using bulk hetero-junction structures, wherein the active layer is spin coated from a mixed solution of donor and acceptor materials. The resulting partially demixed blend structure allows efficient exciton ionization at the large interfacial area, while also maintaining adequate charge transport and extraction through bi-continuous donor and acceptor phases. Organic solar cells are the only PV device that uses molecules to absorb photons and convert them to electric charges without the need of intermolecular transport or electronic excitation. It is also the only variety of solar cell that splits the two functions of light harvesting and charge-carrier transport, whereas conventional PV devices perform both operations simultaneously. This imposes stringent demands upon the optical and electronic properties of the semiconductor, i.e., its band gap and band position, as well as charge-carrier mobility and the recombination time of photogenerated charges, restricting greatly the choice of suitable materials that are able to act as efficient PV converters. Overall research progress on these and other advanced materials, devices, and concepts has been outstanding over the past few years.

This book covers these recent advances in PV materials and their innovative applications. Many material science problems are encountered in understanding existing solar cells and the development of more efficient, less costly, and more stable cells. This important and timely book provides a historical overview, but concentrates primarily on the exciting developments in the last decade. Also, it covers the different Maximum Power Point Tracking (MPPT) controls, which have led to improved speed of response, a better MPP search accuracy, and good control in the presence of perturbations such as sudden variations of illumination and temperature. Furthermore, the optimal design of PV systems based on the two different approaches such as consumed power and economics are discussed. The chapter-wise details of the book are as follows.

Chapter 1 analyzes the global energy trends in terms of achievements, challenges, and outlooks. Also, it deals with X-rays, global energy accessibility, and the role of PV solar cell systems in achieving a global supply of energy with modern energy attributes. Furthermore, the significance of nanotechnology in enhancing the efficiency of PV solar materials is discussed.

Chapter 2 discusses the recent development of utilizing copper paste for solar cell application and its appropriate annealing conditions for better electrical properties. Also, an I-V characteristic of copper paste on the solar cell is summarized. Consequently, 20.7% of the conversion efficiency from the passivated emitter and rear totally diffused structure solar cell confirmed the potential of copper paste as a promising future metallization material.

Chapter 3 discusses the noise characterization in PV devices with ventilation due to solar energy conversion. The experiments are conducted for obtaining parameters such as currents, voltages, temperatures, air velocities, sensible heat capacity, and thermal storage capacity of a PV device with active ventilation. To support the noise wave characterization, signal processing is achieved from a PV device composed of an RC analog signal. The noise characterization is represented with noise filters.

Chapter 4 investigates the field installation of a fixed PV alongside an existing equivalent tracking PV simultaneously monitoring their current and voltage response with changes in solar radiation and ambient temperatures. The efficiency of solar electric systems basically depends on the materials used in making the solar cells regardless of the type of application such as fixed or tracking PV. From the comparative results, it is evident that both systems have a relatively slow electrical response to sunrise, while the performance of a fixed PV system approximates the tracking of PV systems at noon time.

Chapter 5 explores a simplified design and implementation of the impedance-matching stage using a DC/DC buck converter with MPPT for improving the performance of a PV generator. Also, this method enhances the profitability and stability of electrical networks. MPPT control has led to improved speed of response, a better MPP search accuracy, and good control in the presence of perturbations such as sudden variations in illumination and temperature. The system is tested with three different MPPT algorithms such as Hill-Climbing, Perturb & Observe, and Incremental Conductance for different climatic conditions.

Chapter 6 presents a new approach to realize quick MPPT for PV bedded on roads. The proposed method (MPPT with microconverter connected to a short PV string) supports a quick response for the shadow flickers caused by moving objects. The chapter further discusses the PV properties and an MPPT algorithm accelerated by a learning machine using a modal regression on a budget. The proposed method is evaluated by simulation under partial shadow conditions. From the results, it is evident that the proposed method obtains faster tracking than the existing methods such as the MPPT with particle swarm optimization. In conclusion, this method is suitable for electricity generation using the solar panels bedded on roads.

Chapter 7 deals with the optimal design of a PV system based on the two different approaches such as consumed power and economics. The system consists of a PV grid connected through a shunt active filter with MPP tracking. A grid-connected PV system with shunt active filter and different MPPT tracking and designs of DC links for shunt active filters are discussed. Also, it includes the modeling of a PV panel and shunt active filter. The performance of the system is verified using MATLAB/Simulink.

Dr. Natarajan Prabaharan
Assistant Professor
Department of Electrical and Electronics Engineering
SASTRA Deemed University
Tamilnadu, India

Dr. Marc A. Rosen
University of Ontario Institute of Technology
Oshawa, Canada

Dr. Pietro Elia Campana
KTH Royal Institute of Technology, Department of Chemical Engineering
Mälardalen University, Department of Energy, Building and Environment
Mälardalen, Sweden

Section 1

Recent Trends in Photovoltaic Materials

Chapter 1

Solar Energy Conversion and Noise Characterization in Photovoltaic Devices with Ventilation

Himanshu Dehra

Additional information is available at the end of the chapter

http://dx.doi.org/10.5772/intechopen.79706

Abstract

An investigation is performed on solar energy conversion and noise characterization in photovoltaic devices with ventilation. A parallel plate photovoltaic (PV) device was installed with a pair of PV modules, a ventilated air cavity, and an insulating back panel of plywood board filled with polystyrene installed in an outdoor test room. The characterization of noise interference due to power difference of two intensities for composite waves on a PV device is presented. Standard definitions of noise sources, their measurement equations, their units, and their origins under limiting reference conditions are devised. The experiments were conducted for obtaining currents, voltages, temperatures, air velocities, sensible heat capacity, and thermal storage capacity of a PV device with active ventilation through an outdoor test room. Photovoltaic amplification was attained with power output from a potentiometer through the rotation of its circular knob. A parallel plate PV device was studied for its electrical parameters as resistance-capacitance (RC) electrical analog circuit. The effect of inductive and capacitive heating losses was considered in evaluating electrical characteristics of a PV device exposed to solar radiation. Noise filter systems as per noise sources are illustrated with examples. Some examples of noise unit calculations are tabulated based on devised noise measurement equations.

Keywords: solar energy conversion, PV device, photovoltaic amplification, noise characterization, ventilation, solar energy acoustics

1. Introduction

Solar energy conversion occurs at solar cells, and solar intensity of incident solar energy is converted into electric power and waste heat. The photovoltaic devices with ventilation provide means for converting waste heat lost to surrounding environment into useful thermal power. The composite waves are transmitted in photovoltaic (PV) devices due to stresses and oscillations of incident solar and ventilation energy. In this way, solar power intensity is converted into heat, fluid, electricity, light, sound, and fire depending on intensities of its transmitted composite waves in PV devices. This chapter has summarized the concept of noise characterization in PV devices with ventilation due to solar energy conversion. The following sections define and describe noise, its sources and its measurement equations with support of experimental and numerical results of a PV device with ventilation. In order to support the noise wave characterization, signal processing is achieved from a PV device composed of RC analog signal. The noise characterization is exemplified with noise filters. Some noise unit calculations deduced from the devised noise measurement equations are also presented.

1.1. Noise

Noise, defined as "a sensation of unwanted intensity of a wave," is perception of a pollutant and a type of environmental stressor. An environmental stressor such as noise may have detrimental effects on various aspects of health. The unwanted intensity of a wave is noise propagation due to transmission of waves (viz. physical agents) such as light, sound, heat, electricity, fluid, and fire. A unified theory for stresses and oscillations is applicable so as to take into effect of all the physical agents as an environmental stressor on a human body [1]. As per the theory, the stresses developed on a particle due to various forces are classified as fundamental stresses, internal stresses, and external stresses. The fundamental stresses are developed due to the presence of gravitational and electromagnetic forces of a solar system. The internal stresses are developed under the influence of fundamental stresses and are defined by properties and composition of a particle. A theory of noise interference in a wave is deemed based on noise sources and their units [1–5]. The noise sources, their measurement equations and units, are derived from the concept of interference of waves and unified theory of stresses and oscillations [6–8]. The noise filters are classified as per source signal of unwanted frequencies from solar power, electric power, light power, sound power, heat power, fluid power, and fire power [8–10]. This noise concept is also useful for characterization and checking of a human noise behavior [11].

1.2. Noise characterization

The interference of noise arises due to the difference of power of two intensities. The intensity of power for any particle body is a function of the development of various stresses. The phenomenon of acoustic resonance occurs when critical stress level matches with the natural stress level necessary for the oscillation of a particle body [1, 8]. The criteria for generation of acoustic resonance include waves propagated with transmission of light, sound, noise, heat, electricity, fluid, and fire from a particle body. The sensation and perception of noise from light, sound, heat, electricity, fluid, and fire is a physiological response from the sensory organs of a standard (average) human body [8].

The characterization of noise interference due to difference of power of two intensities is conceptualized. The difference of two power intensities is due to the transmission of light, sound heat, electricity, fluid, and fire into a particle body. The sources of noise are classified according to the type of wave of interference, such as light, sound, heat, electricity, fluid, and fire. The criteria for definitions of noise are based on areas of energy stored in a wave, due to interference, speed of wave, and difference of power between two intensities of wave.

2. Sources of noise

The sources of noise are classified according to the type of wave of interference [3]:

Light: In the electromagnetic radiation wavelength band from approximately between 380 and 765 nm, the visual sensation of light is tested by the eye of an observer seeing a radiant energy. The physiological response from an average eye defines the units of light. The sensitivity of human eye is not same in all wavelengths or colors. The contribution of adding daylight is visual sensation in the visible region of the solar energy spectrum.

Sound: In the range of frequencies between 20 and 20,000 Hz, the sound is evaluated due to the presence of fluid pressure energy as a hearing sensation by the ear. The sound units are based on the functional feedback of an average ear. The sensitivity of sound to the whole frequency band is not the same for human ear.

Heat: In the electromagnetic radiation between 0.1 and 100 μm, the heat as a temperature sensation is examined by the human body. The sensation function of temperature defines the units of heat. The temperature sensation function is a measure of coldness and hotness. The comfort zone of temperature is evaluated from functional feedback of a human body which also defines the thermal comfort. The contribution to discomfort of human body is in the ultraviolet region of solar energy spectrum.

Electricity: With passing of direct current or an alternating current, the electricity as a shock sensation is evaluated by skin of an observer due to the electromagnetic energy stored in a conductor which is short circuited by a human body.

Fluid: The fluid as combined ventilation and breathing sensation is evaluated by the amount of fluid passed either externally or internally through a standard (average) human body.

Fire: The exposure of radiant energy and fluid acting on the skin surface of an average human body defines the fire as a sensation of burning.

3. Definitions

The definitions of noise sources are characterized by energy area stored in a wave with its speed and difference due to power intensities of two waves due to the interference [8].

Noise of sol: The difference of power intensities between two solar power systems causes noise of sol (S). The power storage on a unit area per unit time defines the amplitude of a solar energy wave.

The storage of solar power is defined by a solar energy wave pack of unit cross sectional area and of length s, the velocity of light.

Noise of therm: The difference of power intensities between two heat power systems causes noise of therm. The power storage on a unit area per unit time defines the amplitude of a heat wave.

The storage of heat power is defined by the heat energy wave pack of unit cross sectional area and of length s, the velocity of light.

Noise of photons: The difference of power intensities between two lighting power systems causes noise of photons. The power storage on a unit area per unit time defines the amplitude of a light beam.

The storage of light beam is defined by the light beam packet of unit cross sectional area and of length s, the velocity of light.

Noise of electrons: The difference of power intensities between two electrical power systems causes noise of electrons. The power storage on a unit area per unit time defines the amplitude of an electricity wave.

The storage of electrical power is defined by an electricity wave pack of unit cross sectional area and of length s, the velocity of light.

Noise of scattering: The difference of power intensities between two fluid power systems causes noise of scattering. The power storage on a unit area per unit time defines the amplitude of a fluid wave.

The storage of fluid power is defined by the fluid energy wave pack of unit cross sectional area and of length s, the velocity of fluid.

Noise of scattering and lightning: The difference of power intensities between two fire power systems causes noise of scattering and lightning. The power storage on a unit area per unit time defines the amplitude of a fire flash.

The storage of fire power of light is defined by the fire pack of unit cross sectional area and of length s, the velocity of light. The storage of fire power of fluid is defined by the fire pack of unit cross sectional area and of length s, the velocity of fluid.

Noise of elasticity: The difference of power intensities between two sound power systems causes noise of elasticity. The power storage on a unit area per unit time defines the amplitude of a sound wave. The storage of sound power is defined by the sound energy wave pack of unit cross sectional area and of length s, the velocity of sound.

4. Noise measurement equations

The following standard measurement equations are derived and adopted from the standard definitions for sources of noise interference as mentioned in previous sections [8, 9].

http://dx.doi.org/10.5772/intechopen.79706

Noise of sol: For a pack of solar energy wave, the multiplication of solar power storage and the velocity of light give solar power intensity I. On taking logarithm of two intensities of solar power, I_1 and I_2, provides intensity difference. It is mathematically expressed as [8]:

$$Sol = \log(I_1)(I_2)^{-1} \tag{1}$$

But, the logarithmic unit ratio for noise of sol is expressed as *Sol*. The oncisol (oS) is more convenient for solar power systems. The mathematical expression by the following equality gives an oncisol (oS), which is 1/11th unit of a *Sol* [8]:

$$oS = \pm 11 \log(I_1)(I_2)^{-1} \tag{2}$$

Noise of therm: For a pack of heat energy wave, the multiplication of the total power storage and the velocity of light gives heat power intensity I. The pack of solar energy wave and heat energy wave (for the same intensity I) have same energy areas; therefore, their units of noise are the same as *Sol*.

Noise of photons: For a pack of light energy beam, the multiplication of the total power storage and the velocity of light gives light power intensity I. The pack of solar energy wave and light energy beam (for the same intensity I) have same energy areas; therefore, their units of noise are the same as *Sol*.

Noise of electrons: For a pack of electricity wave, the multiplication of the total electrical storage and the velocity of light gives electrical power intensity I. The pack of solar energy wave and electricity wave (for the same intensity I) have same energy areas; therefore, their units of noise are the same as *Sol*.

Noise of scattering: For a pack of fluid energy wave, the multiplication of the total power storage and the velocity of fluid gives fluid power intensity I. On taking logarithm of two intensities of fluid power, I_1 and I_2, provides intensity difference. It is mathematically expressed as [8]:

$$Sip = \log(I_1)(I_2)^{-1} \tag{3}$$

But, the logarithmic unit ratio for noise of scattering is *Sip*. The oncisip (oS) is more convenient for fluid power systems.

The mathematical expression by the following equality gives an oncisip (oS), which is 1/11th unit of a *Sip* [8]:

$$oS = \pm 11 \log(I_1)(I_2)^{-1} \tag{4}$$

For energy area determination for a fluid wave, the water with a specific gravity of 1.0 is the standard fluid considered with a power of ±1 W m^{-2} for a reference intensity I_2.

Noise of scattering and lightning: For a pack of fire wave, the intensity, I, of fire flash with power of light is the multiplication of the total power storage and the velocity of light. Whereas for a pack of fire wave, the intensity, I, of fire flash with power of fluid is the multiplication of the total power storage capacity and the velocity of fluid.

For a noise due to fire flash, the collective effect of scattering and lightning is to be obtained by the superimposition principle.

- For the same intensity I, the pack of solar energy wave and a fire flash with light power has the same energy areas; therefore, their units of noise are the same as *Sol*. The therm power may also be included in fire flash with power of light.
- For the same intensity I, the pack of fluid energy wave and a fire flash with fluid power has the same energy areas; therefore, their units of noise are same as *Sip*. In determining the areas of energy for the case of fluids other than water, a multiplication factor in specific gravity of fluid is to be considered.

Noise of elasticity: For a pack of sound energy wave, the product of the total power storage and the velocity of sound gives sound power intensity I. On taking logarithm of two intensities of sound power, I_1 and I_2, provides intensity difference. It is mathematically expressed as [8]:

$$Bel = \log(I_1)(I_2)^{-1} \tag{5}$$

But, the logarithmic unit ratio for noise of elasticity is *Bel*. The oncibel (oB) is more convenient for sound power systems. The mathematical expression by the following equality gives an oncibel (oB), which is 1/11th unit of a *Bel* [8]:

$$oB = \pm 11\log(I_1)(I_2)^{-1} \tag{6}$$

There are following elaborative points on choosing an *onci* as 1/11th unit of noise [11]:

Reference value used for I_2 is −1 W m^{-2} on positive scale of noise and 1 W m^{-2} on negative scale of noise. In a power cycle, all types of wave form one positive power cycle and one negative power cycle (see **Figure 10**). Positive scale of noise has ten positive units and one negative unit, whereas the negative scale of noise has one positive unit and ten negative units:

i. Each unit of sol, sip, and bel is divided into 11 parts, 1 part is 1/11th unit of noise, and

ii. The base of logarithm used in noise measurement equations is 11.

The reference value of I_2 is −1 W m^{-2} with I_1 on positive scale of noise, should be taken with negative noise measurement expression (see Eqs. (2), (4) and (6)); therefore, it gives positive values of noise.

The reference value of I_2 is 1 W m^{-2} with I_1 on negative scale of noise, should be taken with positive noise measurement expression (see Eqs. (2), (4) and (6)); therefore, it gives negative values of noise.

Some noise unit calculation examples are illustrated later in this chapter.

4.1. Limiting conditions and reference values

Table 1 has summarized the units of noise and their limiting conditions [3].

Reference*	Noise scales and limiting conditions		
$I_2 = \pm 1$ W m^{-2}	Noise of sol	Noise of scattering	Noise of elasticity
Units	Sol	Sip	Bel
$I_1 = 1$ W m^{-2}	No positive solar energy	No positive fluid energy	No positive sound energy
$I_1 = 1+ \rightarrow 0$ W m^{-2}	Decreasing solar energy	Decreasing fluid energy	Decreasing sound energy
I_1 = +ve	Increasing solar energy	Increasing fluid energy	Increasing sound energy
$I_1 = -1$ W m^{-2}	Negative solar energy	Negative fluid energy	Negative sound energy
	Darkness	Low pressure	Inaudible range
I_1 = −ve	Darkness increasing, distance from point source of light increasing	Low pressure increasing, vacuum approaching	Inaudible range increasing, vacuum approaching
$I_1 = -1+ \rightarrow 0$ W m^{-2}	Negative solar energy	Negative fluid energy	Negative sound energy
	Decreasing darkness	Decreasing low pressure	Decreasing inaudible range

*Reference value of $I_2 = \pm 1$ W m^{-2} signifies the limiting condition with areas of noise interference approaching to zero.

Table 1. Noise under limiting conditions.

5. Noise filter systems

The criteria for definitions of filters for noise filtering are based on the areas of energy stored in a wave due to the noise interference, speed of wave, and difference of power between two intensities of wave [8]. The filtered noise signals are considered from systems of solar power, electric power, light power, sound power, heat power, fluid power, and fire power. The noise filters as per sources of noise are defined as follows [8]:

Filter for noise of sol: This filter is used to filter noise due to the difference of intensities of power between two solar power systems. Example: window curtain, window blind, wall, and sunglasses.

Filter for noise of therm: This filter is used to filter noise due to the difference of intensities of power between two heat power systems. Example: house, insulation, clothing, and furnace.

Filter for noise of photons: This filter is used to filter noise due to the difference of intensities of power between two lighting systems. Example: 3-D vision of any object, electric bulb, television, computer, and LCD screen laptop.

Filter for noise of electrons: This filter is used to filter noise due to the difference of intensities of power between two electrical power systems. Example: AM/FM radio clock with ear phones, telephone instrument with ear phones, and CD audio player with ear phones.

Filter for noise of scattering: This filter is used to filter noise due to the difference of intensities of power between two fluid power systems. Example: electric fan, pump, motor vehicle, river stream, rain, and tap water.

Filter for noise of scattering and lightning: This filter is used to filter noise due to the difference of intensities of power between two fire power systems. Example: lighter, matchstick, gas stove, locomotive engine, and thunderbolt.

Filter for noise of elasticity: This filter is used to filter noise due to the difference of intensities of power between two sound power systems. Example: human vocal chords, organ pipe, thunderbolt, and drum beats.

5.1. Some examples of noise filters

Some examples of noise filters are enumerated as under [10]:

Human voice production: The example of phonetics of filtering sound of a human speech is illustrated. The human speech is synthesized due to the development of stresses at vocal folds. The smoothening of the sound is a function of its amplitude and its shape of oscillations at vocal tract of a human being. The vocal tract is a resonant cavity wall with sound energy stored in oscillations of its vocal folds. The vocal apparatus showing the mechanism of synthesis of human speech is illustrated in **Figure 1**.

An airflow window with a photovoltaic solar wall: The filtering of solar energy is illustrated through an example of an airflow window attached with a shading device. An airflow window is fixed with a movable roller blind to control the transmission of daylight as well as the amount of solar heat. The bottom portion of photovoltaic solar wall is used for controlling the amount of air ventilation along with the generation of solar electric power. The example is illustrated in **Figure 2**.

Psychrometric air conditioner: An elementary air conditioner for summer comfort conditioning consists of a cooling coil, a cooling fluid with a filter. The schematic of operation of a psychrometric air conditioner is illustrated in **Figure 3**.

Telephone line: The impedance of a telephone line is composed of distributed resistance, capacitance, and inductance. The impedance of telephone line is proportional to the insulation, loop length, and whether the wire is buried, aerial or bare parallel wires strung on

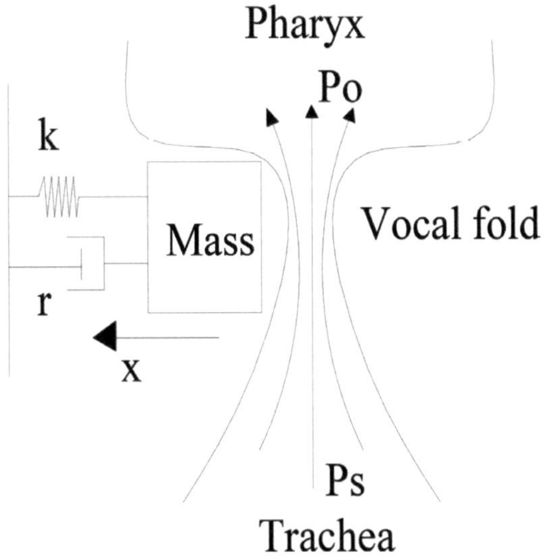

Figure 1. A human vocal mechanism.

http://dx.doi.org/10.5772/intechopen.79706

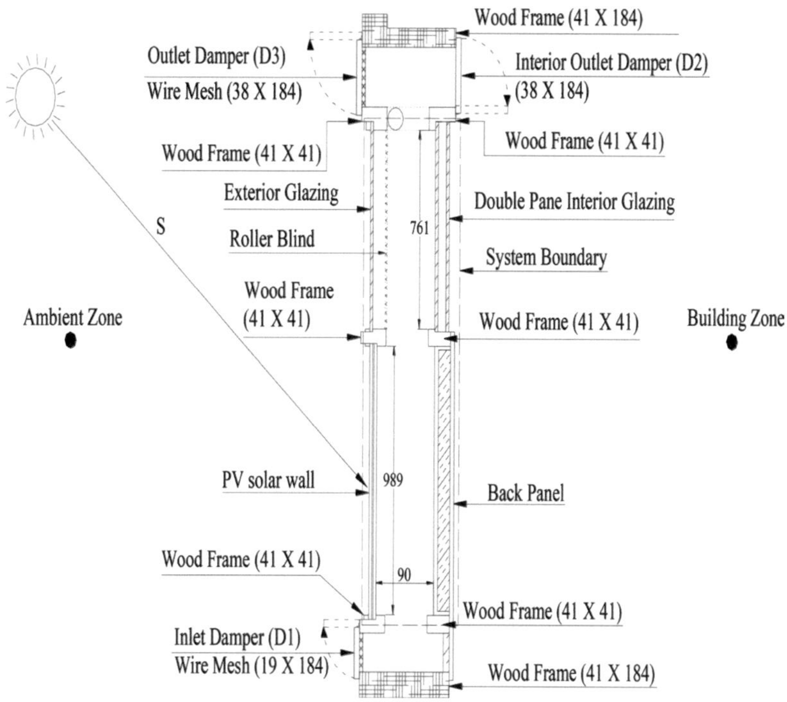

Figure 2. An airflow window with a photovoltaic solar wall (dimensions shown are in mm).

telephone pole. A telephone line is usually supplied with a 48 VDC from the telephone exchange. The schematic of operation of a telephone line with telephone instrument is illustrated in **Figure 4**.

Fire and smoke detection system: A fire detection system consists of a control system with interconnected alarms, smoke, and heat detectors. A fire detector is a device which is used for presetting an alarm at a particular temperature. A smoke detector is a device which is used for presetting an alarm when a certain percentage of smoke accumulates. The photovoltaic cell activates the smoke alarm only if it senses requisite obscuration of light over a unit area with control from BMS. The schematic of various components for fire detection system is illustrated in **Figure 5**.

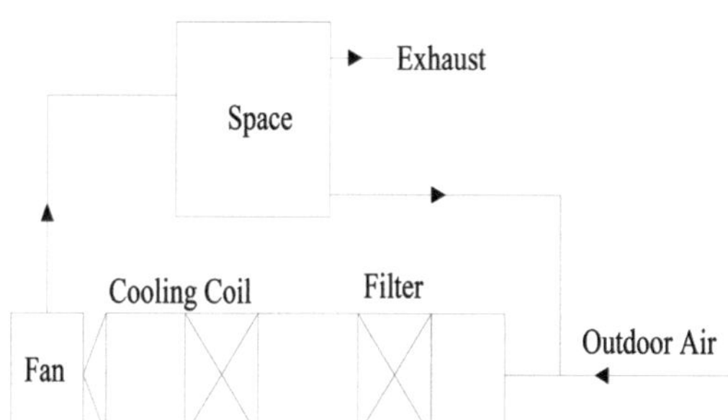

Figure 3. A psychrometric air conditioner.

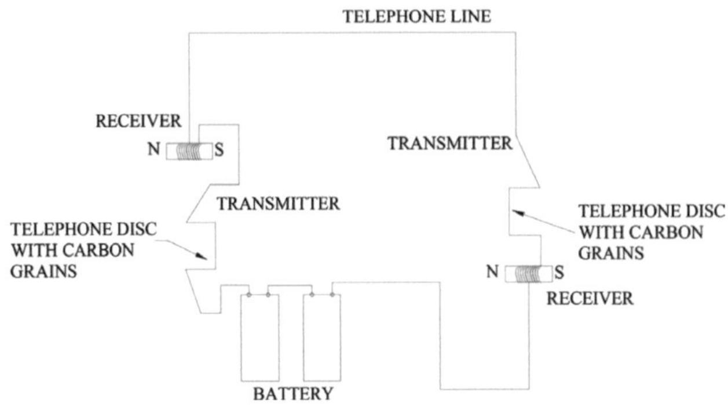

Figure 4. Operation of a telephone line.

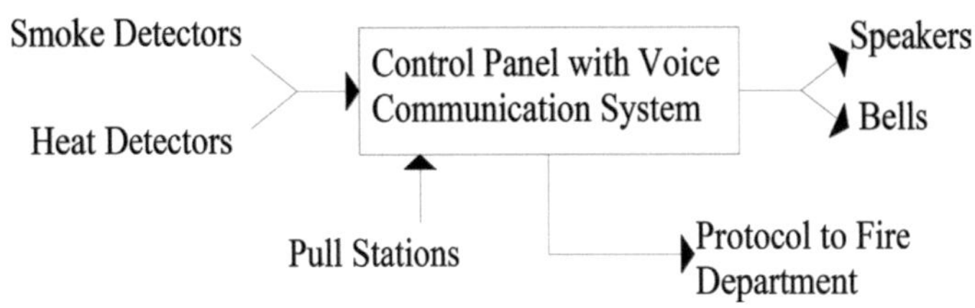

Figure 5. A fire detection system.

6. Experimental and numerical results

The full scale experimental setup for a parallel plate photovoltaic device connected to a potentiometer was installed in an outdoor room facility located at Concordia University, Montréal, Québec, Canada [12–23]. The schematic of the experimental setup is illustrated in **Figure 6**. An amplifier was built with a pair of photovoltaic (PV) modules forming a parallel plate channel with a plywood board and connected to a potentiometer. A potentiometer, a wire-wound variable resistor of up to 50 Ω, was a wire-wound circular coil with a sliding knob contact [23]. It was used to vary electrical resistance across connected PV modules without interrupting the current. The characteristics of a parallel plate photovoltaic device connected to a potentiometer were established by varying electrical resistance with the rotation of knob of a potentiometer. The current-voltage measurements were obtained for determining electric power output with a series electrical circuit connection of a pair of vertically inclined PV modules installed on a wooden frame.

The temperatures were measured as a function of volume of a parallel plate photovoltaic device. The nonlinear thermal results include measurements of temperatures for PV modules, insulating panel, and ventilated air column in the wooden frame. The air velocities were developed in the ventilated air column for the transportation of heat both as a measure of buoyancy and fan induced ventilation. The electrical measurement results of currents, voltages, and power with varying electrical resistance of potentiometer are presented in **Table 2**. The thermal measurement results of temperatures of various components of PV device, ambient air and room air temperatures, air velocities and solar intensities are presented in **Table 3**. The location and nomenclature of sensors are presented in **Table 4**. The results of the power output from a potentiometer with the rotation of circular knob are illustrated in **Figure 7**.

http://dx.doi.org/10.5772/intechopen.79706

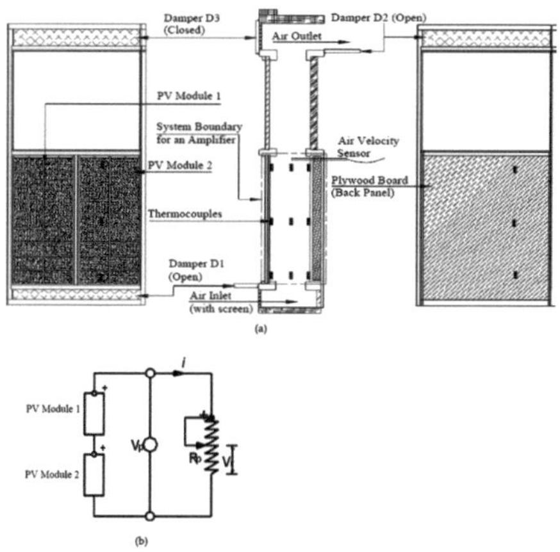

Figure 6. Schematic of experimental setup for a parallel plate photovoltaic device connected to a potentiometer: (a) location of sensors; and (b) electrical circuit diagram.

A simulation model for the prediction of temperature distributions varying with the volume of a parallel plate photovoltaic device was developed. The model was used to predict the temperature distributions at pre-defined locations in PV module, plywood board, and air flowing through a parallel plate channel through the walls of vertically inclined photovoltaic modules. The model results of the temperature plots are illustrated in **Figure 8(a)–(c)**. Some noise unit examples for an air duct exposed to solar radiation are illustrated in **Tables 5–8**. These tables have shown noise unit calculations on positive scale of noise.

6.1. Photovoltaic amplification

The phenomenon of photovoltaic amplification is observed from the graphs of **Figures 7** and **8** [21]. The gain in steady state electrical and thermal functions for a photovoltaic device is a factor of its volume or resistance. This operational characteristic is similar to the operation of a loudspeaker. The electrical analog is used to describe the resonance phenomenon for

Rotation	Volts	Amps	Watts	Rotation	Volts	Amps	Watts	Rotation	Volts	Amps	Watts
240°	18.7	—	—	83°	16.3	0.935	15.23	30°	9.7	1.577	15.21
239°	16.5	0.331	5.461	75°	16.0	1.014	16.26	27°	9.0	1.587	14.33
201°	17.4	0.414	7.195	69°	15.8	1.100	17.38	21°	7.1	1.583	11.24
185°	17.5	0.454	7.940	64°	15.5	1.165	18.04	18°	6.2	1.573	9.831
162°	17.3	0.513	8.885	55°	15.0	1.302	19.53	17°	5.7	1.578	9.026
150°	17.18	0.550	9.449	50°	14.5	1.386	20.05	12°	3.9	1.567	6.257
142°	17.19	0.582	10.00	43°	13.2	1.503	19.79	10°	3.2	1.553	4.840
128°	17.1	0.640	10.93	42°	13.1	1.493	19.49	1.5°	0.5	1.593	0.807
107°	16.8	0.750	12.51	37°	11.9	1.536	18.26	1°	0.3	1.59	0.426
89°	16.4	0.884	14.45	32°	10.5	1.567	16.42	0°	—	1.643	—

Table 2. Sample electrical measurement results of a PV device with varying resistance of potentiometer.

Run no.	S (W m^{-2})	Ep (W)	To (°C)	Ts (°C)	v (m s^{-1})	T_p(b) (°C)	T_p(m) (°C)	T_p(t) (°C)	T_b(b) (°C)	T_b(m) (°C)	T_b(t) (°C)	T_a(b) (°C)	T_a(m) (°C)	T_a(t) (°C)
1	697.5	31.0	13.5	22.1	0.451	34.5	33.01	36.2	20.2	24.4	27.6	18.7	19.3	22.5
2	725.4	31.1	15.9	22.9	0.362	32.5	33.3	35.7	20.2	23.9	29.1	18.3	19.7	23.3

Table 3. Sample thermo-fluid measurement results.

Locations shown in Figure 1	T_p(b) (°C)	T_p(m) (°C)	T_p(t) (°C)	T_b(b) (°C)	T_b(m) (°C)	T_b(t) (°C)	T_a(b) (°C)	T_a(m) (°C)	T_a(t) (°C)	Air velocity sensor
y (cm)	15	55	94	15	55	94	15	55	94	99
Z (cm)	60	60	60	60	60	60	60	60	60	60
X (mm)	6.2	6.2	6.2	96.2	96.2	96.2	51.2	51.2	51.2	51.2

Note: x is horizontal; y is vertical; and z is adjacent 3rd axis of x-y plane.

Table 4. Nomenclature and location of sensors.

equivalent mechanical, hydraulic, and thermal systems of a parallel plate photovoltaic device connected to a potentiometer. **Figure 9(a)** and **(b)** shows series and parallel cases of L-C-R arrangement of resonance, respectively.

6.2. Signal processing: electrical parameters for a PV device

The sinusoidal steady-state response was applied in performing the analysis of the parallel plate PV device circuit, because of the advantage of representing a periodic function in terms of a sinusoidal exponential function. Electrical analog RC circuit parameters of a parallel plate PV device are enumerated as under [2, 13]:

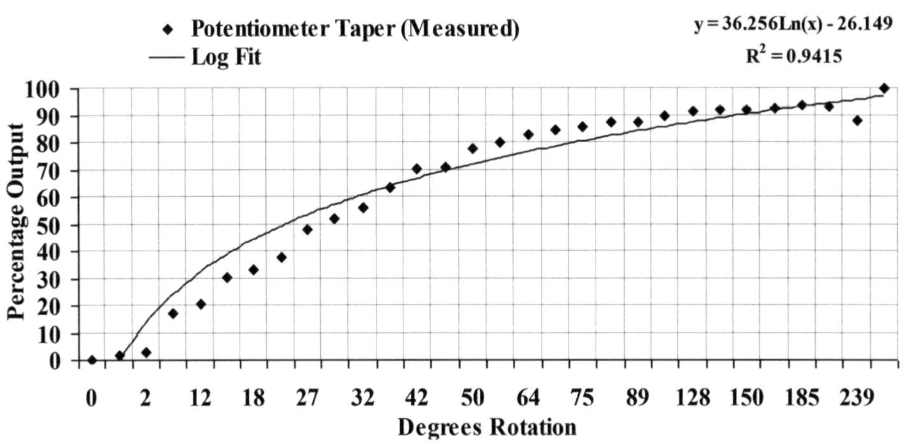

Figure 7. Potentiometer taper (measured) with percentage voltage output.

http://dx.doi.org/10.5772/intechopen.79706

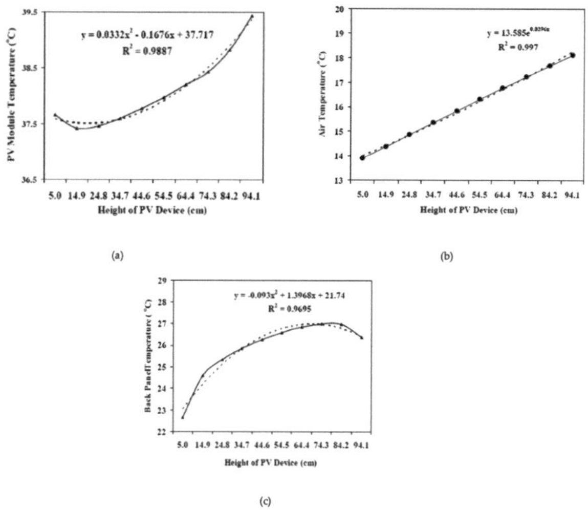

Figure 8. Temperature plots with height of a PV device: (a) PV module; (b) air; and (c) plywood board.

Solar irradiation (W m^{-2})	Air temperature difference (ΔT) (°C)	Noise of Sol oS (oncisol)
450	15.50	28
550	18.90	28.93
650	22.40	29.7
750	25.90	30.36
850	29.40	30.91

Table 5. Temperature difference and noise of sol with solar irradiation (air velocity: 0.75 m s^{-1}).

Air velocity (m s^{-1})	Fluid power (W m^{-2})	Air temperature difference (ΔT) (°C)	Noise of scattering oS (oncisip)
1.35	47.62	15.28	17.72
1.05	37.0	18.22	16.50
0.75	26.45	22.40	15.02
0.45	15.87	28.15	12.65
0.15	05.29	29.80	07.64

Table 6. Temperature difference and noise of scattering with air velocity (S = 650 W m^{-2}).

Capacitance: The capacitance of a parallel plate PV device with air as a dielectric medium was calculated to be 91.2 picofarads.

(ΔT) °C	Mass flow rate (kg s⁻¹)	Thermal power (W m⁻²)	Noise of therm oS (oncisol)	(ΔT) (°C)	Mass flow rate (kg s⁻¹)	Thermal power (W m⁻²)	Noise of therm oS (oncisol)
15.50	0.01376	71.09	19.5602	15.28	0.0231	117.65	21.868
18.90	0.01275	80.325	20.119	18.22	0.0171	103.85	21.296
22.40	0.0120	89.6	20.614	22.40	0.0120	89.6	20.614
25.90	0.0115	99.2833	21.043	28.15	8.1×10^{-3}	76.0	19.866
29.40	0.0111	108.78	21.505	29.80	6.2×10^{-3}	61.59	18.898

Table 7. Mass flow rate and noise of therm with (ΔT) (°C).

Air velocity ($m\ s^{-1}$)	Fluid power ($W\ m^{-2}$)	Noise of scattering oS (oncisip)	Sound pressure ($N\ m^{-2}$)	Sound power intensity ($W\ m^{-2}$)	Noise of elasticity oB (oncibel)
1.35	47.62	17.72	557.5	752.7	30.36
1.05	37.0	16.50	433.65	455.33	28.05
0.75	26.45	15.02	309.75	232.31	24.97
0.45	15.87	12.65	185.85	83.63	20.24
0.15	05.29	07.64	61.94	09.29	10.12

Table 8. Noise of elasticity with air particle velocity (impedance $Z_0 = 413\ N\ s\ m^{-3}$ at 20°C).

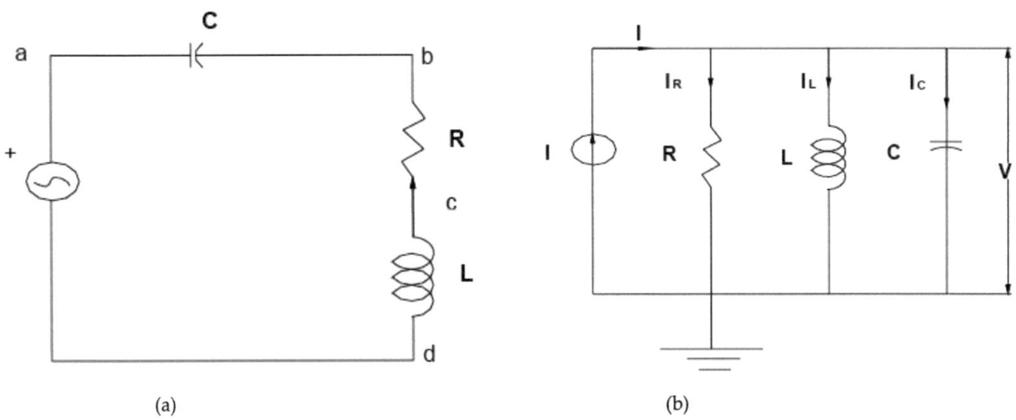

Figure 9. (a) L-C-R series arrangement of resonance and (b) L-C-R parallel arrangement of resonance.

Resistance: The electrical resistances of various components were calculated as: glass coated PV modules were approximated as 5.3 kΩ, air was approximated as 1200 MΩ, and plywood board was approximated as 26.5 Tera Ω. The total equivalent electrical resistance of a parallel plate PV solar wall device was approximated as 5.3 kΩ.

Time constant: The time constant, which is a product of resistance and capacitance, was calculated to be: 0.5 μs. The frequency with this time constant was calculated to be 2 MHz.

Capacitive reactance: The capacitive reactance was calculated to be 872.5 Ω.

Impedance: The impedance of the circuit was calculated to be 5.4 kΩ.

The phase angle θ: The phase angle between capacitance and reactance was calculated to be 9°.

The phasor representation:

$$Z = 5.300 - j0.8725 = 5.4\ k\Omega \ \llcorner\ -9^{\circ}.$$

Capacitive heating: The joule law gives instantaneous power absorbed by the capacitive impedance which is converted into heat. The heat capacities under critical operation of buoyancy-induced ventilation were calculated to be 59.6, 0.755, and 510.7 for PV module, air, and plywood board, respectively. The total average value of joule heating for the parallel plate PV device was calculated to be 571 kJ.

Induction losses: The induction losses due to the thermal storage effect in the parallel plate PV device were calculated to be 15.9 kJ.

Power factor: The power factor was calculated to be cosθ = 0.911 lag.

Current function ($i^2(t)$): Using the current function, $i^2(t) = I_m^2 \sin^2(\omega t + \theta)$, the effective (root mean square) value of current was calculated to be 10.4 amps, and the maximum value of current was calculated to be 14.71 amps.

Voltage function: The voltage function is defined as per the sine wave: $v = V_m \sin(\omega t)$. The effective value of the voltage was calculated to be 60.4 V, and the maximum value of the voltage was calculated to be 85.42 V.

Power function: The instantaneous power is given by the expression [2]:

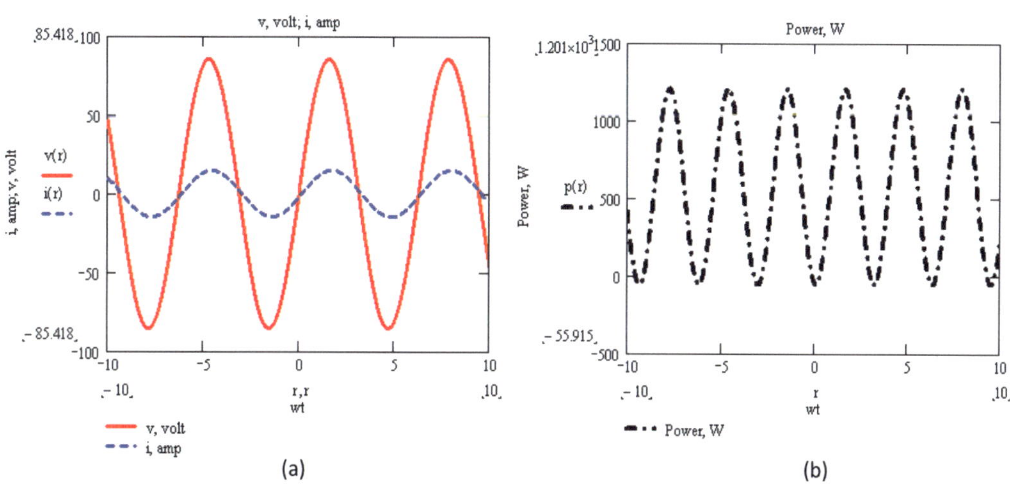

Figure 10. Time diagrams: (a) voltage and current and (b) power in a RC circuit amplifier.

$$p(t) = \frac{Vm.Im}{2}.\cos(\theta) - \frac{Vm.Im}{2}.\cos(2\omega - \theta) \quad (7)$$

The plots: The time diagram for current and voltage is plotted in **Figure 10(a)**. The time diagram for power is plotted in **Figure 10(b)**.

Power transfer: **Figure 10** shows that the instantaneous power is negative whenever the voltage and current are of opposite sign. However, as is illustrated in **Figure 10** that positive area of p(t) energy exceeds the negative area. Therefore, the average power is finite. Since the angle, θ, is small between current and voltage, the negative area of energy becomes very small. During the first quarter cycle (from 0° to 90°), the applied voltage rises from slightly negative value to a maximum, and the capacitor is receiving a charge. The power curve is positive during this period and represents energy stored in the capacitor. From 90° to 180°, the applied voltage is falling from maximum to slightly negative value, and the capacitor is discharging. The corresponding power curve is negative and represents energy returned to the circuit during this interval. The third quarter cycle represents a period of charging the capacitor, and the fourth quarter represents a discharge period. The induction losses are due to the thermal storage amount to 1.5% in comparison to the capacitive heating. Thus, induction losses cannot be avoided in any electrical circuit, but can be minimized.

7. Discussion

The following composite waves are generated due to the development of stresses and oscillations on a PV solar wall device with incident short wavelength electro-magnetic waves [1]: (i) due to the connected external electrical load and transmission of electrical energy wave; (ii) due to the exchange of viscous dissipation with air and the propagation of heat waves at longer wavelength; (iii) due to the thermal stress generation with propagation of heat waves, and elastic waves are transmitted in a PV solar wall device; (iv) due to the combination of stress development with heat waves, elastic waves, and applied external source of energy, and the fluid surface waves are propagated; and (v) due to the climate particle oscillations of wind and fan induced pressure, and applied external waves are propagated; in the absence of wind and fan pressure, thermosyphon-based oscillations are propagated due to the thermal buoyancy [18, 23].

Such as in an organ pipe, the sound waves can transmit with the combination of applied external source of energy and fluid surface waves. From the surrounding environment due to air-borne sound transmission, the sound waves are propagated. Due to various stresses and oscillations acting on a static particle body, the transmission of composite waves is generated. With the action of composite forces acting on a PV solar wall device, the developed stresses are classified as: (i) fundamental; (ii) internal; and (iii) external. Due to the presence of electromagnetic and gravitational forces of a solar system, the fundamental stresses are generated. Under the influence of fundamental stresses, internal stresses are generated with characterization of composition properties at chemical, atomic, and molecular level. With application of external source of energy such

http://dx.doi.org/10.5772/intechopen.79706

as fan-based force used for active air ventilation in PV solar wall device, the external stresses are generated. With stress development due to the periodic force of expansion/compression, cooling/heating and night/day, the oscillations are assumed to be generated on the PV solar wall device. On a PV solar wall device, climate particle oscillations due to wind force are also transmitted. Due to the superposition of composite waves, the fluctuating forces are generated.

Resonance: The parallel and series cases of LCR circuit resonance are briefed here [8, 10]. With the aid of presented modeling and experimental data, the cases of resonance are visualized. For elastic waves transmission, the inductance force exists due to the mass of the mechanical system. The capacitance force exists due to the heat storage capacities of PV solar wall device (PV modules, air, and polystyrene filled plywood board). The polystyrene filled plywood board is vulnerable to heat stress of fire as soon as heat waves propagated with frequency matching with its latent heat of vaporization is achieved. Due to the thermal and fluid resistance in energy storage elements of the PV solar wall device, equivalent electrical analog resistance is developed. The parallel case of LCR resonance happens with fluid surface waves (RC) and heat waves (RC) in conjunction with inductance (L) due to the mass of PV solar wall device and resistance (R) due to temperatures of ambient air and ground surface. The series case of LCR resonance occurs with propagation of elastic waves of a PV solar wall device.

8. Conclusion

A study on solar energy conversion and noise characterization in photovoltaic devices with ventilation is performed. The noise interference and characterization as per speed of a composite wave are presented. The sources of noise waves (sun, light, sound, heat, electricity, fluid, and fire) are described depending on their speed of noise interference. Noise measurement equations and their units are coined. The power systems are classified as per source signals of solar power, electric power, light power, sound power, heat power, fluid power, and fire power. The noise filters for filtering noise from power systems are defined with examples. The experimental results along with results of the simulation model for noise filtering for a PV device are presented. Some noise unit examples for an air duct exposed to solar radiation are illustrated. A phenomenon of photovoltaic amplification for a pair of photovoltaic modules connected to a potentiometer is explained. The time plots of power function were used to support and devise noise measurement expressions and noise characterization in a power system as per speed of a wave.

Author details

Himanshu Dehra

Address all correspondence to: anshu_dehra@hotmail.com

Egis Group, Gurugram, Haryana, India

References

[1] Dehra H. A unified theory for stresses and oscillations. In: Proc. CAA Conf., Montréal 2007 Canada, Canadian Acoustics. Vol. 35(3). 2007. pp. 132-133

[2] Dehra H. Power transfer and inductance in a star connected 3-phase RC circuit amplifier. In: Proc. AIChE 2008 Spring Meeting, New Orleans, LA, USA. Session 96a. 2008. 7 p

[3] Dehra H. The noise scales and their units. In: Proc. CAA Conf., Vancouver 2008 Canada, Canadian Acoustics. Vol. 36(3). 2008. pp. 78-79

[4] Dehra H. A benchmark solution for interference of noise waves. In: Proc. AIChE Spring 2009; Tampa, FL, USA; April 26-30, Session 67c. 2009. 4 p

[5] Dehra H. Solar energy absorbers. In: Manyala R editor. Chapter 6 in Solar Collectors and Panels, Theory and Applications. Intech Publication; 2010. pp. 111-134

[6] Dehra H. A theory of acoustics in solar energy. Natural Resources. 2013;**4(1A)**:116-120

[7] Dehra H. A slide rule for noise measurement. In: 10th International Conference on Sustainable Energy Technologies (SET 2011); Istanbul, Turkey; September 4-7. 2011. 5 p

[8] Dehra H. A novel theory of psychoacoustics on noise sources, noise measurements and noise filters. In: Proc. NoiseCon16 Conf.; 13-15 June; Providence, Rhode Island, USA. 2016. pp. 933-942

[9] Dehra H. On sources and measurement units of noise. In: Proc. International Conference on Innovation, Management and Industrial Engineering (IMIE 2016); 05-07 August 2016; Kurume, Fukuoka, Japan; 2016. pp. 219-227. ISSN: 2412-0170

[10] Dehra H. Acoustic filters. In: Romano VA, Duval AS, editors. Chapter 5 in Ventilation: Types, Standards and Problems. New York, USA: Nova Science Publishers; 2011. 245 pp. ISBN: 978-1-62618-281-3

[11] Dehra H. A paradigm for characterization and checking of a human noise behavior. In: WASET Proceedings of 19th International Conference on Psychological and Brain Sciences; May 11-12, 2017; Montréal, Canada. pp. 317-325. http://waset.org/publications/10007615

[12] Dehra H. Photovoltaic solar wall: 2-D numerical modeling and experimental testing under fan induced hybrid ventilation. In: Proceedings of the IEEE International Conference on Energy Efficient Technologies for Sustainability, April 7-8. IEEE Xplore, ISBN: 978-1-4673-9925-8; 2016. pp. 668-676

[13] Dehra H. A multi-parametric PV solar wall device. In: Proceedings of IEEE International Conference on Power, Control, Signals and Instrumentation Engineering (ICPCSI-2017), 978-1-5386-0814-2/17/$31.00 ©2017 IEEE, Chennai, India on 21-22 Sep 2017

[14] Dehra H. Characterization of noise in power systems. In: Proceedings of IEEE International Conference on Power Energy, Environment & Intelligent Control (PEEIC2018), 978-1-5386-2341-1/18/$31.00 ©2018 IEEE, Greater Noida, India on April 13-14, 2018. pp. 321-330

[15] Dehra H. A guide for signal processing of sensors and transducers. In: Proc. AIChE 2009 Spring Meeting; Tampa, FL, USA. 2009. Session 6b

[16] Dehra H. A numerical and experimental study for generation of electric and thermal power with photovoltaic modules embedded in building façade [submitted/unpublished Ph.D. thesis]. Montréal, Québec, Canada: Department of Building, Civil and Environmental Engineering, Concordia University; August 2004

[17] Dehra H. The effect of heat and thermal storage capacities of photovoltaic duct wall on co-generation of electric and thermal power. In: AIChE. 2007 Spring Meeting; April 22-26; Houston, Texas, USA. Session 36a. 2007. 9 p

[18] Dehra H. The entropy matrix generated exergy model for a photovoltaic heat exchanger under critical operating conditions. International Journal of Exergy. 2008;**5**(2):132-149

[19] Dehra H. A two dimensional thermal network model for a photovoltaic solar wall. Solar Energy. 2009;**83**(**11**):1933-1942

[20] Dehra H. Electrical and thermal characteristics of a photovoltaic solar wall with passive and active ventilation through a room. International Journal of Energy and Power Engineering. 2017;**11**(5):514-522. http://waset.org/publications/10007024

[21] Dehra H. An investigation on energy performance assessment of a photovoltaic solar wall under buoyancy-induced and fan-assisted ventilation system. Applied Energy. 2017;**191**(1):55-74

[22] Dehra H. A combined solar photovoltaic distributed energy source appliance. Natural Resources. 2011;**2**:75-86

[23] Dehra H. A mathematical model of a solar air thermosyphon integrated with building envelope. International Journal of Thermal Sciences. 2016;**102**:210-227

Chapter 2

Conductive Copper Paste for Crystalline Silicon Solar Cells

Sang Hee Lee and Soo Hong Lee

Additional information is available at the end of the chapter

http://dx.doi.org/10.5772/intechopen.78604

Abstract

In photovoltaic industries, the main technique of metallization is screen printing with silver pastes due to its simple and quick process. However, the expensive price of silver paste is one of the barriers to the production of low-cost solar cells. Therefore, the most focused target in photovoltaic research is the decreasing consumption of silver paste or substitute silver for other materials. As a proper candidate, copper has been researched by many institutes and companies since it has a similar conductivity with silver even though the price is inexpensive. To apply copper as a contact for solar cells, the plating technique has been actively researched. However, copper paste, which was mainly developed for integrated circuit applications, has been recently researched. Mostly, copper paste was developed for the low-temperature annealing process since copper tends to oxidize easily. On the other hand, firing type copper paste was also developed by coating copper particles with a barrier layer. This chapter discusses recent development of copper paste for the application of solar cells and its appropriate annealing conditions for better electrical properties. Also, the light I-V characteristics of copper paste on the solar cells in other research papers are summarized as well.

Keywords: copper paste, oxidation barrier coating, curing, silicon heterojunction solar cells, passivated busbar

1. Introduction

In photovoltaic industries, screen printing is the most dominant metallization technique for silicon-based solar cell fabrication as it is quick and simple. As a material of front contact, silver is the favorable metal since it has high conductivity [1] and is chemically inactive. However, screen printing with silver paste is the most expensive portion in cell production after the silicon material cost [2]. Therefore, reducing the amount of silver consumption per cell or replacing silver to other metal materials is a significant research area.

According to the international technology roadmap for photovoltaic (ITRPV) published in 2016, silver consumption per cell will decrease until 40 mg/cell in 2026 with developments of pastes and screens, which is around 40% lower than now (95 mg/cell) [3]. On the other hand, substituting silver for copper has been actively researched since the cost of copper is cheaper than silver (approximately a 50 times) and has a similar conductivity (silver: 1.6 μΩ-cm, copper: 1.7 μΩ-cm) [4, 5]. In order to share new information and go over the technical limitations, workshops for the metallization of crystalline silicon solar cells have been organized since the first workshop in Utrecht, Netherlands, in 2008 [6].

Researches concerning copper contact mainly have been carried out by the plating technique due to its various advantages, such as high aspect ratio and low contact resistance, which result in a high-efficiency solar cell over 21% [7–12]. Meanwhile, the application of screen-printable copper paste on solar cells has been studied as it can be easily applied to the established cell production line. In the case of the copper paste, copper particles cannot be deposited directly on the emitter, because the copper atoms have fast diffusion velocity and acts as a deep-level impurity in the crystalline silicon solar cell [13–18]. The copper atoms in the silicon produce generation and recombination centers and degrade the minority carrier lifetime of the crystalline silicon solar cells [11, 19–21]. Accordingly, most of the copper pastes on the solar cells were printed above the passivation layer as a busbar, which is called "passivated busbars", while the silver paste fingers contacted the silicon. **Figure 1** shows the fingers and a busbar of the solar cell that are printed by silver paste. Similar to the finger, the role of the busbar on the solar cell is a collection of charge carriers generated by incident light in the absorb layer. The busbar is also connected to the soldered ribbon to extract carriers out of the device. To connect a busbar with a ribbon, the busbar should be printed with similar width of the ribbon which is usually 1.5 mm on the commercial type of solar cells. Accordingly, researchers have tried to apply copper paste only for the busbar since most of the silver paste usage is for the busbar, while the fingers were still printed by silver paste or deposited by the plating technique.

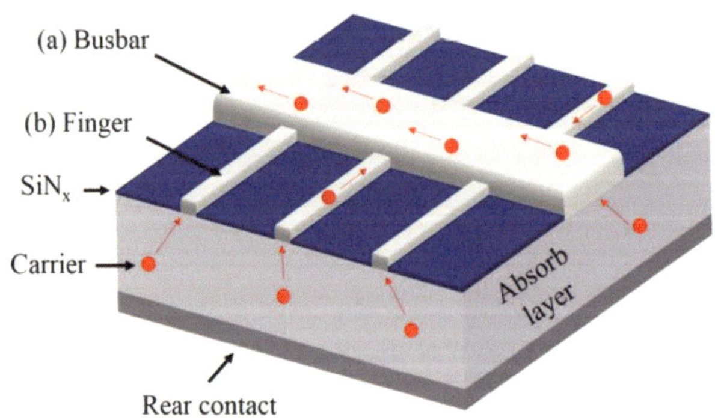

Figure 1. Carrier collection by the screen-printed silver (a) busbar and (b) finger [26].

Nonetheless, it is possible that the copper paste can be in direct contact with the silicon if the copper particles are coated with barrier layers in order to prevent copper from diffusing into the silicon. Another issue of copper in the application to the paste form is that copper tends to oxidize easily during thermal treatment [13, 22–25]. Since copper oxide shows an electrically nonconductive characteristic, it will increase the series resistance in the solar cells. Due to these reasons, copper paste has been continuously researched by several institutes and companies in order to overcome such issues. Section 2 deals with research trends of the copper paste components and promising coating techniques of copper powder for better reliability. Afterwards, Section 3 discusses appropriate curing conditions of polymer-based copper paste and the results of copper paste application to the silicon solar cells.

2. Copper paste developments for the crystalline silicon solar cells

In the 1990s, copper paste was researched for the application of integrated circuits, such as print circuit boards, because copper has a high electrical conductivity, a high thermal conductivity, excellent solderability, and a low electron migration [27–29]. With the increase in circuit density, properties of copper paste needed to be improved. The researched topics were mainly focused on optimizing the size of metal particles to enhance the density and printability of paste [30, 31]. Recently, copper paste has been developed for the application of the crystalline silicon solar cells as the low-cost front contact.

2.1. Structure of metallized solar cells with screen-printed pastes

Copper paste is generally compared to silver paste since it is a dominant material for the front metallization of the crystalline silicon solar cell. In order to apply copper paste to the solar cells, the properties of copper paste, such as printability and solderability, need to have similar or better characteristics than silver paste. Electronic pastes are generally composed of conductor metal (Ag, Au, Pd, Cu, etc.), glass frits, and organic vehicle [32, 33]. One of the important components of the conventional silver paste for the front contact of the crystalline silicon solar cell is glass frits. In case of the crystalline silicon solar cells based on the silver paste, the dielectric layer, which is usually silicon nitride (SiN_x), is fired-through above 600°C and the silver particles contact the emitter (**Figure 2(a)**).

On the other hand, if the copper paste has the same process as the silver paste, the diffused copper can adversely effect on the characteristics of the solar cell as we mentioned earlier. Thus, for applying copper paste to the crystalline silicon solar cell, copper particles in the paste need to be coated by barrier layers. In this case, the copper paste can be fired at a similar temperature range as the silver paste and contact the emitter (**Figure 2(b)**). Otherwise, the contact should be formed without penetrating the SiN_x layer (**Figure 2(c)**) by applying the curing type copper paste which does not need the glass-frits components for the fired-through contact.

Silicon heterojunction (SHJ) solar cells have typically a low process temperature limit (~250°C) because high-temperature annealing processes can degrade the passivation of the

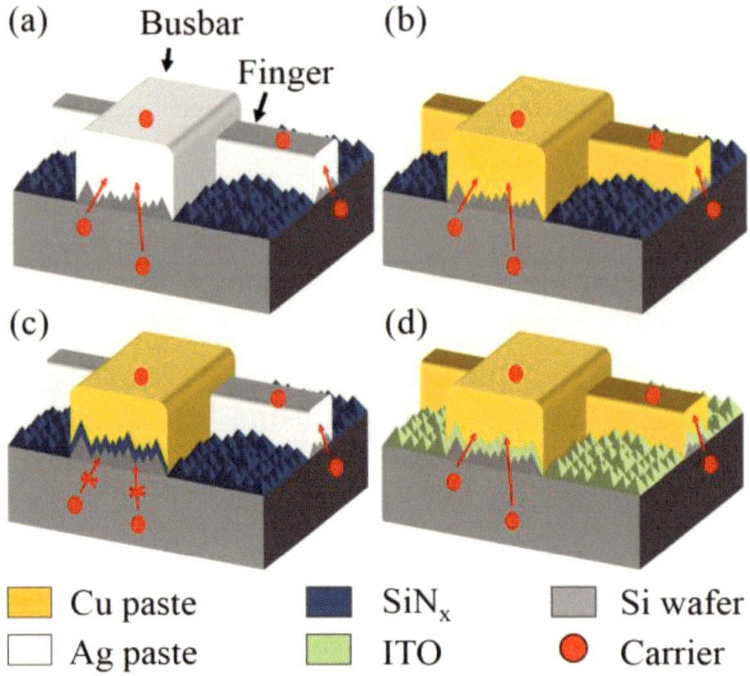

Figure 2. (a) Conventional silicon solar cell contact with silver paste, (b) contact with firing-type copper paste, (c) contact with curing-type copper paste, and (d) contact with curing-type copper paste on the SHJ solar cell [26].

hydrogenated amorphous silicon (a-Si:H) due to the hydrogen effusion during the annealing [34]. For this reason, the curing-type copper paste, where low temperature is generally required, is beneficial to the SHJ solar cells. Also, if copper paste is printed on the indium tin oxide (ITO) layer of the SHJ solar cell (**Figure 2(d)**), ITO can act as a diffusion barrier for preventing copper diffusion [35]. The next section discusses detail components of the copper pastes for the application of the solar cells by categorizing the annealing temperature of the paste.

2.2. Copper paste for high-temperature annealing (firing type)

In 2011, a copper paste that is chemically and metallurgically similar to conventional silver paste was developed by Applied Materials, Inc. [36]. The copper paste can be fired through a SiN_x layer and the metal particles directly contact silicon (**Figure 2(b)**). The main components and possible materials of the invented copper paste are listed in **Table 1**. The invented technique involves copper-containing particles being encapsulated by additional layers of metal and alloys to restrict oxidation and diffusion of copper during the firing. For improving the oxidation resistance of copper, alloying copper with other metals (Ti, Mg, Al, Pd, Ag, Ni, Cr, and Zr) has been researched [37–40]. The Cu-Ag alloy is estimated as the best materials for improving oxidation resistance with only a slight reduction in electrical conductivity [41]. The paste of this group also uses doped copper or copper alloys rather than pure copper particles,

Components	Materials	Purposes
Metal powder	Doped copper (aluminum, magnesium, etc.), copper alloys (CuSn, CuAg, CuNi, CuZn, etc.)	To have conductive property
Polymer resin	Ethylcellulose	To enable printing of copper-containing particles
Glass frits	Lead oxide (PbO_x), silicon oxide (SiO_2), alumina (Al_2O_3), boron trioxide (B_2O_3), zirconia (ZrO_2), zinc oxide (ZnO), bismuth oxide (Bi_2O_3), strontium oxide (SrO), titanium oxide (TiO_2), and lanthanum oxide (La_2O_3)	To pass through a passivation layer and promotes adhesion to the substrate
Solvents	α-Terpineol, toluene, ethanol	To tune viscosity characteristics

Table 1. Components and possible materials of the copper paste for high-temperature annealing.

because alloying elements in copper reduce the contact with oxygen. Generally, the polymer resin acts as a binder to enable printing of the encapsulated copper-containing particles and is typically removed during the firing by oxidation.

Figure 3 shows three levels of encapsulation for preventing copper particles from oxidation and diffusion. Simply, the copper-containing particle can be coated by oxidation barrier layers. Also, a metallization barrier layer can be used under the oxidation barrier since the oxidation layer can form an alloy with the inside material. Moreover, a diffusion barrier can directly surround the copper-containing particle for a more perfect encapsulation. The possible materials for the encapsulation layer are listed in **Table 2**.

2.3. Copper paste for low-temperature annealing (curing type)

In order to create a solderable surface on the ITO of the SHJ solar cells, polymer-based silver pastes were commonly used in the solar cell industry, because silver has a low contact resistivity on ITO and low line resistances. However, reactions between polymer and solder flux during the annealing result in a "solder leaching" problem. If the screen-printed paste is dissolved in the solder material due to the solder leaching, it leads to low adhesion and high contact resistivity between paste and solder material.

Using copper instead of silver, as a metal powder in the polymer-based paste, is a good solution in order to overcome the issue of solderability, because copper produces a comparable solderability and resistance at a much lower price [42]. For these reasons, polymer-based copper paste for low-temperature annealing has been researched as a promising product in the future with the fact that the SHJ solar cells have become common in the PV industry [43–46]. From now on, the components and properties of the curing-type copper pastes from some groups will be discussed.

2.3.1. Dow Corning

Dow Corning reported papers and patents about a curing-type copper paste and the applicable solar cell structures [47–51]. The curing is referred to as the hardening of polymer materials

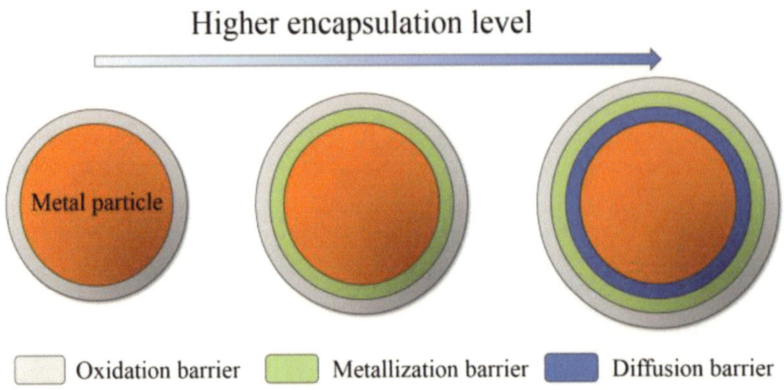

Figure 3. Cross-sectional views of encapsulated copper-containing particles with single and multi-barrier layer.

Encapsulation layer	Oxidation barrier	Metallization barrier	Diffusion barrier
Possible materials	Silver (Ag), nickel (Ni), and zinc (Zn)	Nickel (Ni), titanium (Ti), titanium nitride (TiN), tungsten (W), titanium-tungsten (TiW), cobalt (Co), tungsten doped cobalt (Co:W), molybdenum (Mo), tantalum (Ta), and chromium (Cr)	

Table 2. The possible materials of each encapsulation layer for copper-containing particles.

by cross-linking polymer chains that can be processed by heating at a low temperature under 300°C. The copper paste consists of metal powder, solder powder (lower melting temperature than that of the metal powder), a polymer, a solvent, a cross-linking agent, and additives. The solder powder comprises at least one of a tin-bismuth (SnBi) alloy, a tin-silver (SnAg) alloy, or combinations of them. The polymer and the carboxylated polymer are made of an epoxy resin and an acrylic polymer, respectively. The cross-linking agent (or catalyst) can be chosen from carboxylated polymers, dimer fatty acids, and trimer fatty acids. Among the dimer fatty acid, dicarboxylic acid and monocarboxylic acid are useful for fluxing the metal powder and cross-linking the polymer. Moreover, a solvent and an adhesion promoter can be included as additive components.

This copper paste is used to form a busbar of the conventional crystalline silicon solar cell without a fired-through process. **Figure 4** shows that the printed busbar has a brown-red color due to the copper particles. Afterwards, the color of the busbar changes to gray after the curing process because the copper particles are coated by the solder. The cells with the copper busbar have a higher front surface minority carrier lifetime than the cells with the silver fired-through busbar since the covered area under the busbar is fully passivated. The detail characteristics will be mentioned in Section 3.

2.3.2. National Institute of Advanced Industrial and Scientific Technology (AIST)

A research group in the AIST also reported a similar concept of the copper paste as the Dow Corning's copper paste. Their copper paste, which is called "copper-alloy paste," is composed

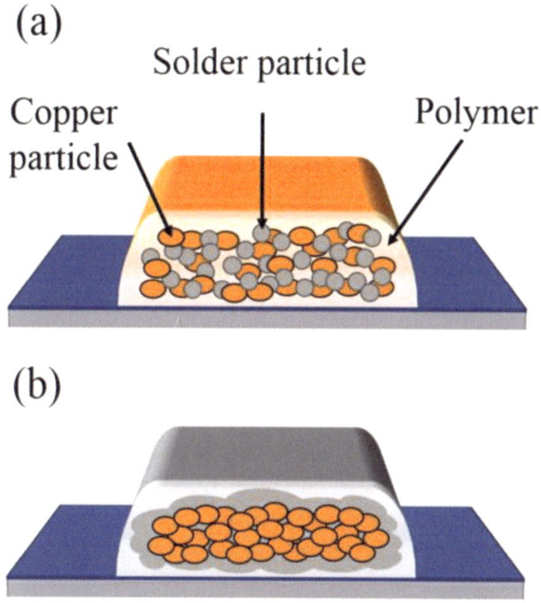

Figure 4. Dow Corning's screen-printable copper paste: (a) after printing and (b) after curing [26].

of conductive metal particles, low melting point alloy (LMPA), thermosetting polymer, and solvent [52]. During the curing process, the molten LMPA particles form alloy with the copper particles and surround the copper particles to prevent oxidation. In particular, the LMPA allows the curing process to set the temperature below 200°C without any reductive conditions unlike the conventional silver paste. The result of the differential scanning calorimetry (DSC) shows that the melting point of the LMPA is 143°C. The peak of the DSC graph is very sharp since the LMPA had a nano-level uniformity. Moreover, the copper-alloy paste shows better self-leveling and resolution than the conventional silver paste after the screen-printing process on a textured silicon wafer.

Also, the copper paste from this group shows decent reliability after printing as a busbar on p-type crystalline silicon [53]. The samples were tested by the damp heat test (DHT) and thermal cycling test (TCT) before and after the encapsulation with the "sandwich" structure (glass/EVA/cell/EVA/backsheet) according to the IEC61215 standards. The results of both DHT and TCT show degradation less than 5% of the initial values in all parameters (V_{oc}, J_{sc}, FF, $P_{ma}x$, etc.) before and after encapsulation. Although the surface of the copper electrode without encapsulation is oxidized after the DHT test, the copper oxide layer acts as a semi-passivation layer that postpones inner oxidation. Moreover, the copper particles in their paste do not diffuse into the silicon even after an hour of annealing at 400°C due to the polymer barrier layer.

2.3.3. Samsung Electro-Mechanics Co., Ltd.

The invented copper paste was focused on the nano-particle size copper powder, especially for substrates (such as a transparent conductive oxide (TCO), a polymer, a glass plate, and a printed circuit board), which have difficulties in applying high-temperature processes [54, 55].

The average particle size of copper is around 150 nm, and the surfaces of the copper particles are coated with a capping material which can be fatty acid or fatty amine. The nano-size copper powder is used either solely as a metal powder or with different sizes of copper particles, such as a flake powder and a spherical powder. The flake powder has a particle size of 1–20 μm and the spherical powder has a 0.1–5-μm particle size. When the nano-powder is mixed with other types of powder, it first dissolves during the annealing and then helps to connect between the larger copper particles. Because of this nano-size effect, this copper paste can enhance conductivity. The detailed candidates for binders and additives are also presented in the patent. Consequently, the copper particle at 150-nm size decreases the annealing temperature of the paste and makes it possible to form electrodes at a low temperature of 200°C.

2.3.4. *Institute of Nuclear Energy Research (INER)*

Recently, the INER reported an antioxidant copper paste [56, 57]. The antioxidant copper nanoparticles are synthesized by a wet chemical reduction process which requires copper hydroxide ($Cu(OH)_2$), polyvinylpyrrolidone (PVP), and ascorbic acid. Afterwards, the antioxidant copper nanoparticles are transferred to the paste form and printed onto the ITO layer of SHJ solar cells, followed by low-temperature annealing (<300°C). Compared to commercial silver pastes as a reference, this copper paste shows a twofold increase in sheet resistance (~30 mΩ/sq) on the 16 μm of printed films. However, the duration of copper paste annealing is 1/12 of that of silver paste. Also, reserving samples for 180 days without strict oxygen protection shows no peaks of oxide impurities after XRD characterization, which means that the copper film is relatively stable against oxidation at least at an X-ray detection level.

2.4. Promising techniques for high performance of copper paste

2.4.1. *Coating of copper powder with nano-silica*

In order to apply copper on conductive paste, it requires high-purity crystalline non-agglomerated copper powder, which is free from surface oxidation [58, 59]. Using silica as a coating material of copper powder can enhance colloidal properties and functions by using rational core-shell shapes [60]. Dong et al. coated nano-copper powder with nano-silica by using a sol–gel process to improve the dispersion of the glass in the paste, the density of films, and the bonding behavior between the film and the substrate [61]. The printed films by using the copper paste after sintering at 910°C show no significant change in the density of the surface morphology and sheet resistance with the contents of silica from 0.5 to 2 wt%. However, the bonding between the film and the substrate improves with 2 wt% of silica contents in copper powder. The reason is that the proper amount of silica contents can induce the capillary effects and surface sorption effects which is beneficial to bond the film closely on the substrate. The properties of silica-coated copper powder will be able to improve the bonding of the high-temperature annealing copper paste on the silicon wafers.

2.4.2. *Coating of copper powder with cobalt-catalyzed carbon nanofibers*

Even though the properties of polymer-based copper paste have been improved by many research, it is still difficult to achieve high conductivity and reliability as silver paste due to the relatively low conductivity of the polymers [62–66]. In addition, using nanoscale copper particles for decreasing curing temperature also have issues of powder production step, such as controlling the size of particles [67], low oxidation resistance of particles [68, 69], and cost-effectiveness [70]. For this reason, the development of copper particles, which are coated by a carbon-based material, has been interested by many researchers, because carbon shells can act as the shields to protect the copper particles from oxidation [71–79]. In addition, there have been studies for the development of copper paste or ink, which do not require inert atmosphere and lower temperature, but they still have challenges to overcome [80–83].

In order to improve oxidation resistance of copper particles and make curing process possible in air, Ohnishi et al. coated copper-cobalt alloy particles with cobalt-catalyzed carbon nanofibers (CNFs) which is called "hybrid copper particles (HCuP)" [84]. The paste, which is made by the sea urchin-shaped copper particles, shows great reliability of resistivity even after a DHT test. The good electrical properties of this copper paste might come from an anti-oxidation effect of CNFs. Moreover, the cobalt nano-precipitates on the surface of the particles can be regarded as a conductive path. This approach possibly can improve the reliability of copper pastes by curing without strictly controlled inert atmosphere.

3. Application of copper paste on crystalline silicon solar cells

3.1. Curing conditions of copper paste for high electrical properties

Low-temperature annealing paste generally contains polymer as a component. Accordingly, the properties of these pastes highly depend on the polymerization quality during the curing process. The curing process is carried out at a lower-temperature range than the firing process, which is generally used for the conventional silver paste. *Rehm thermal systems GmbH* and *Fraunhofer Institute for Ceramic Technologies and Systems* reported the effect of curing conditions on properties of the electrode which is printed with the polymer-based copper paste [22, 42, 62, 85]. By using an inert inline drying system, they show that curing with a high nitrogen atmosphere and temperature at 200°C can significantly decrease the resistance of copper paste electrode [42].

The main reason of the resistance reduction is that the cross-linking reactions of polymer chains are sensitive to the oxygen concentration, because the oxygen disturbs the linking process between the polymer chains. **Figure 5** shows the reactions of degradation of the polymer chain which frequently occurs in a high oxygen atmosphere. Oxygen easily reacts with most organic radicals which form "peroxidic radicals" [86]. The peroxidic radicals can suffer the polymerization reactions or the chain processes. As one of the most frequent reactions, an oxidative degradation breaks the polymer chains by initiating the decomposition of the peroxidic radicals. If the polymerization process is carried out in this circumstance, oxygen presence will decrease the cross-linking yield of polymers in the pastes. Therefore, the inert curing atmosphere with low concentration of the oxygen is significant in order to make an intensified polymerization and increase the compression of the metal particles. The restrained oxidation of the metal particles can also be a possible reason.

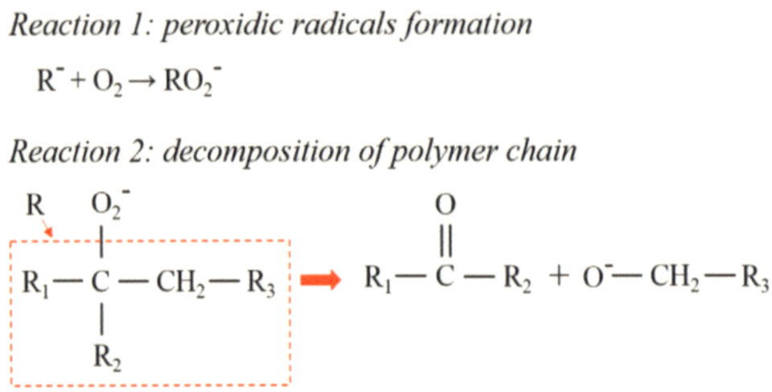

Figure 5. Reactions of polymer chain decomposition by oxygen.

In case of the heat transfer method, a radiation method is more beneficial for the lower resistance of electrode than a convection method [62]. Moreover, the minimum resistance and decent adhesion can be obtained by increasing the processing time [22]. Consequently, this group confirmed that the polymer-based copper paste, which was annealed by the inert curing, can improve conductivity and mechanical stability of the polymer-based copper paste by achieving 19.96% efficiency with the SHJ solar cell, even though the fill factor (FF) is still lower than that of silver paste-printed cells.

3.2. Potential of copper paste on the silicon solar cells as passivated busbars

Some research groups have tried to apply their own copper paste to solar cells. The copper pastes were printed as passivated busbars that required forming busbars and fingers separately. As **Figure 4** shows, fingers only electrically contact silicon by using either fired-through silver paste (**Figure 6(b)**) or the plating of Ni/Cu/Ag metal stack after the laser ablation opening of SiN_x layer (**Figure 6(a)**). Afterwards, the busbar is printed on the SiN_x layer and partially contacts the fingers followed by a curing process under 250°C. Since the busbars do not directly contact the silicon, recombination region under the busbars is removed. Light I-V performances of the solar cells with copper paste busbar are summarized in **Table 3**. On the reference cells, either the screen-printed silver paste contact or the Ni/Cu/Ag-plated contact was wholly used for the busbars and fingers. Generally, the reduced recombination on the front side contributes to an increase open circuit voltage (V_{oc}) compared to the cells without passivated busbar.

Dow Corning and IMEC evaluated characteristics of various cell structures by applying their own low-temperature (~250°C) copper paste for the passivated busbars. The research results show a slight increase of V_{oc} (0.3 mV) with an industrial level passivated emitter solar cell (PESC) by reducing the recombination region under the busbars. Compared to the conventional silver paste solar cell, the passivated copper busbar solar cell has a lower average fill factor (FF) due to the higher lateral resistivity of the copper busbar. However, the busbar resistivity does not have an effect on the FF in the module level performance since most of lateral current flows through the conductive soldered tab.

This group also evaluated combinations of printable conductive copper paste with higher efficiency solar cell structures, such as passivated emitter and rear cell (PERC) and passivated

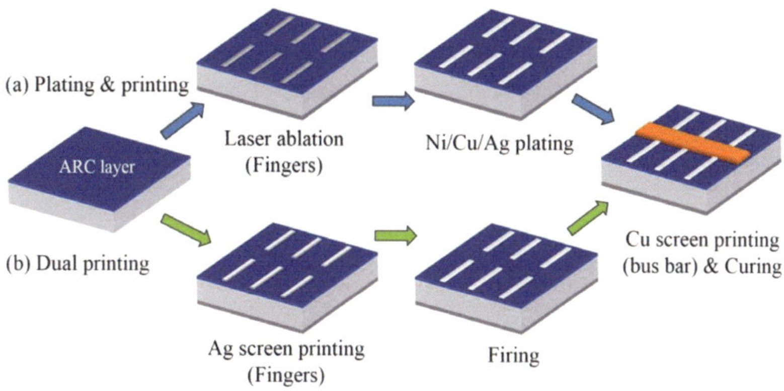

Figure 6. Front metallization process flows for the passivated copper busbar: (a) plating and printing, and (b) dual printing [26].

Institute	Year	Cell type	Ref. contact	η [%] (gain)	V_{oc} [mV] (gain)	J_{sc} [mA/cm^2]	FF [%] (gain)	Annealing temp.	Ref.
Dow Corning	2014	p-PESC (SE)	SP*-Ag	18.8 (−0.07)	640.4 (+0.3)	37.2	78.9 (−0.5)	~250°C	[47]
Dow Corning	2015	p-PERC	Plated-Ni/Cu/Ag	20.4 (+0.1)	667.5 (+6.1)	38.7	79.0 (+0.3)	~250°C	[48]
Dow Corning	2015	n-PERT	Plated-Ni/Cu/Ag	20.7 (0)	663.3 (+4.9)	39.2	79.5 (0)	~250°C	[48]
Meiji Univ.	2015	n-PERT (bifacial)	SP-Ag	20.5 (+0.7)	659.0 (+3.0)	40.8	76.3 (+2.0)	—	[87]
Tokyo Univ.	2012	p-PESC (mc-Si)	SP-Ag	16.2 (+0.1)	617.0 (−2.0)	34.3	76.3 (+0.1)	<200°C	[88]

*SP: screen printed.

Table 3. Performance of various solar cell structures with the passivated copper busbar and gains compared to their reference contact.

emitter and rear totally diffused (PERT). By applying a copper paste busbar with the plating and printing process as depicted in **Figure 6(a)**, both structures improved 6.1 mV, 4.9 mV of V_{oc}, respectively, and the PERC structure especially had a 0.1% higher median conversion efficiency than the reference group. Also, the FF of the passivated busbar cells had increased since the laser ablation and the nickel silicide decreased the shunt resistance of entirely plated cells. In the case of the current density, the passivated busbar cell had a slightly lower value even though the series resistance of both the printed busbar and the plated busbar almost had no difference since the plated busbar had a higher aspect ratio (fine line width).

Nakamura et al. at the Meiji University applied copper paste on the n-type bifacial PERT cell and successfully obtained over 20% efficiency by preceding V_{oc} and FF of the silver-printed cell. Also, Yoshiba et al. at the Tokyo University compared the I-V performances on the multi-crystalline silicon solar cell by printing a low melting point alloy (LMPA) copper paste. Although V_{oc} was decreased, copper-printed cell had a 0.1% absolute efficiency gain due to the higher FF. In most of the experiments for confirming applicability of copper paste on solar cell fabrication, the results show the possibility of reduction of metallization cost and cell performance improvement by replacing the standard silver-printed electrode.

4. Summary and outlook

In this chapter, a detailed overview of the copper paste developments for the solar cell application has been presented. The main issues of developing copper paste are prohibition of the oxidation of copper during annealing and the diffusion into the silicon substrate. In case of the glass-frit-based copper paste (firing type), the copper particles are coated with metal or alloy layers to prevent the diffusion and the oxidation. However, the firing-type copper paste still has a higher possibility of diffusion than the polymer-based copper paste (curing type) since the copper particle comes in direct contact with the silicon. In case of the curing-type copper paste, the diffusion of copper particles is well blocked since the surrounding polymer acted as a barrier layer. Also, the oxidation of copper can be prevented by the polymer shield or using antioxidant copper particles. Moreover, DHT and TCT of the copper paste confirm the reliability on the solar cells with a small amount of degradation (<5%).

For further improvement of the copper paste properties, recently reported coating materials and techniques for the copper powder have been introduced. In case of the nano-silica coating on copper powder, the bonding strength of paste on the substrate was improved by promoting capillary effects and surface sorption effects. Also, the air-curable hybrid copper particles, which were coated by cobalt-catalyzed-CNFs, lead to a great resistance reliability of the printed copper paste.

With respect to the curing conditions, the experimental results revealed that the inert atmosphere helps to form a denser copper electrode by restricting the contact between the polymers and the oxygen. Thereby, the compressed copper particles due to the intensified polymerization decrease the resistivity of the printed copper film. However, the inert curing condition requires a great deal of nitrogen gas for purging oxygen in the furnace. At the industrial level, the nitrogen consumption can adversely affect the manufacturing cost of the solar cells. Therefore, the optimum curing process for less consumption of nitrogen gas and inexpensive coating technique of copper particles need to be further developed.

To date, polymer-based copper paste showed a high potential with 20.7% conversion efficiency by applying it to the n-PERT structure solar cells. As well as the result of the PERT structure, the copper paste application to SHJ solar cells has a higher potential because the ITO layer acts as a diffusion barrier to prevent copper at a low curing temperature. Also, the use of copper paste, as the passivated busbars, requires an additional printing, and the annealing step after the silver fingers and aluminum rear contact are formed. Accordingly, the SHJ solar cells are more profitable for the application of copper paste from an economic point of view.

http://dx.doi.org/10.5772/intechopen.78604

Acknowledgements

This work was supported by the Korea Institute of Energy Technology Evaluation and Planning (KETEP) and the Ministry of Trade, Industry & Energy (MOTIE) of the Republic of Korea (No. 20173010012940) and by the Ministry of Trade, Industry, and Energy, Korea Evaluation Institute of Industrial Technology (KEIT) (No. 10043793).

Author details

Sang Hee Lee and Soo Hong Lee*

*Address all correspondence to: shl@sejong.ac.kr

Department of Electronics Engineering, Green Strategic Energy Research Institute, Sejong University, Seoul, Korea

References

[1] Park K, Seo D, Lee J. Conductivity of silver paste prepared from nanoparticles. Colloids and Surfaces A: Physicochemical and Engineering Aspects. 2008;**313**:351-354

[2] Powell DM, Fu R, Horowitz K, Basore PA, Woodhouse M, Buonassisi T. The capital intensity of photovoltaics manufacturing: Barrier to scale and opportunity for innovation. Energy & Environmental Science. 2015;**8**:3395-3408

[3] International Technology Roadmap for Photovoltaic (ITRPV). Available from: http://www.itrpv.net [Accessed: 2016-Mar]

[4] London Bullion Market Asscociation (LBMA). Available from: http://www.lbma.org.uk/home [Accessed: 2017-Mar]

[5] London Metal Exchange. Available from: https://www.lme.com/metals/non-ferrous/nickel [Accessed: 2017-Mar]

[6] Workshop on Metallization for Crystalline Silicon Solar Cells. Available from: http://www.metallizationworkshop.info

[7] Horzel J, Bay N, Passig M, Sieber M, Burschik J, Kühnlein H, Brand A, Letize A, Lee B, Weber D, Böhme R. Low cost metallization based on Ni/Cu plating enabling high efficiency industrial solar cell. In: Proceeding of the 29th European Photovoltaic Solar Energy Conference and Exhibition; 2014; Amsterdam, Netherland. pp. 507-512

[8] Bay N, Horzel J, Passig M, Sieber M, Burschik J, Kühnlein H, Bartsch J, Brand A, Mondon A, Eberlein D, Völker C, Gutscher S, Letize A, Lee B, Weber D, Böhme R. Reliable contact formation for industrial solar cell by laser ablation and Ni/Cu plating. In: Proceedings

of the 29th European Photovoltaic Solar Energy Conference and Exhibition; 2014; Amsterdam, Netherland. pp. 1272-1276

[9] Hernández JL, Adachi D, Yoshikawa K, Schroos D, Assche EV, Feltrin A, Valckx N, Menou N, Poortmans J, Yoshimi M, Uto T, Uzu H, Hino M, Kawasaki H, Kanematsu M, Nakano K, Mishima R, Kuchiyama T, Koizumi G, Allebé C, Terashita T, Hiraishi M, Nakanishi N, Yamamoto K. High efficiency copper electroplated heterojunction solar cells. In: Proceedings of the 27th European Photovoltaic Solar Energy Conference and Exhibition; 2012; Frankfurt, Germany. pp. 655-656

[10] Papet P, Hermans J, Söderström T, Cucinelli M, Andreetta L, Bätzner D, Frammelsberger W, Lachenal D, Meixenberger J, Legradic B, Strahm B, Wahli G, Brok W, Geissbühler J, Tomasi A, Ballif C, Vetter E, Leu S. Heterojunction solar cell with electroplated Ni/Cu front electrode. In: Proceedings of the 28th European Photovoltaic Solar Energy Conference; 2013; Paris, France. pp. 1976-1979

[11] Bartsch J, Mondon A, Kamp M, Kraft A, Wendling M, Mehling M, Wehkamp N, Jawaid M, Lorenz A, Clement F. Copper as conducting layer in the front side metallization of crystalline silicon solar cells—Challenges, processes and characterization. Proceedings of the 2nd Workshop on Metallization for Crystalline Silicon Solar Cells—Status, Trends and New Directions, Constance, Germany, Konstanz, Germany; 2010. pp. 32-37

[12] Metz A, Adler D, Bagus S, Blanke H, Bothar M, Brouwer E, Dauwe S, Dressler K, Droessler R, Droste T. Industrial high performance crystalline silicon solar cells and modules based on rear surface passivation technology. Solar Energy Materials and Solar Cells. 2014;**120**:417-425

[13] Istratov AA, Weber ER. Physics of copper in silicon. Journal of The Electrochemical Society. 2002;**149**:G21-G30

[14] Brotherton S, Ayres J, Gill A, Van Kesteren H, Greidanus F. Deep levels of copper in silicon. Journal of Applied Physics. 1987;**62**:1826-1832

[15] Kaimao C, Zhongan W. Deep levels related to copper in silicon. Journal of Electronics (China). 1988;**5**:285-293

[16] Aboelfotoh M, Svensson B. Copper passivation of boron in silicon and boron reactivation kinetics. Physical Review B. 1991;**44**:12742

[17] Murarka SP, Hymes SW. Copper metallization for ULSL and beyond. Critical Reviews in Solid State and Materials Sciences. 1995;**20**:87-124

[18] Murarka SP. Advanced materials for future interconnections of the future need and strategy: Invited lecture. Microelectronic Engineering. 1997;**37**:29-37

[19] Schibli E, Milnes A. Deep impurities in silicon. Materials Science and Engineering. 1967;**2**:173-180

[20] Weber ER. Transition metals in silicon. Applied Physics A: Materials Science & Processing. 1983;**30**:1-22

[21] Ur Rehman A, Lee SH. Crystalline silicon solar cells with nickel/copper contacts. In: Solar Cells—New Approaches and Reviews. InTech; 2015

[22] Clement C, Bell H, Vogg F, Rebenklau L, Gierth P, Partsch U. Inert drying system for copper paste application in PV. Energy Procedia. 2013;**38**:423-429

[23] Mimura K, Lim J-W, Isshiki M, Zhu Y, Jiang Q. Brief review of oxidation kinetics of copper at 350 C to 1050 C. Metallurgical and Materials Transactions A. 2006;**37**:1231-1237

[24] Xu X, Luo X, Zhuang H, Li W, Zhang B. Electroless silver coating on fine copper powder and its effects on oxidation resistance. Materials Letters. 2003;**57**:3987-3991

[25] Jeong S, Woo K, Kim D, Lim S, Kim JS, Shin H, Xia Y, Moon J. Controlling the thickness of the surface oxide layer on cu nanoparticles for the fabrication of conductive structures by ink-jet printing. Advanced Functional Materials. 2008;**18**:679-686

[26] Lee SH, Lee DW, Lee SH. Review of conductive copper paste for c-Si solar cells. Korean Journal of Metals and Materials. 2017;**55**:637-644

[27] Wu S. Preparation of ultra-fine copper powder and its lead-free conductive thick film. Materials Letters. 2007;**61**:3526-3530

[28] Huang C-Y, Sheen SR. Synthesis of nanocrystalline and monodispersed copper particles of uniform spherical shape. Materials Letters. 1997;**30**:357-361

[29] Sinha A, Das SK, Kumar TV, Rao V, Ramachandrarao P. Synthesis of nanosized copper powder by an aqueous route. Journal of Materials Synthesis and Processing. 1999;**7**:373-377

[30] Ogawa T, Asai T, Itoh O, Hasegawa M, Ikegami A, Atoh K, Kobayashi T. New thick-film copper paste for ultra-fine-line circuits. IEEE Transactions on Components, Hybrids, and Manufacturing Technology. 1989;**12**:397-401

[31] Taylor BE, Tsuchiya M, Okabe C. Thick Film Copper Paste Composition. Google Patents; 1992

[32] Brown O, Sridharan S. Lead-free and Cadmium-free Conductive Copper Thick Film Pastes. Google Patents; 2004

[33] Brown O, Sridharan S. Lead-free and Cadmium-free Conductive Copper Thick Film Pastes. Google Patents; 2006

[34] De Wolf S, Kondo M. Boron-doped a-Si: H/ c-Si interface passivation: Degradation mechanism. Applied Physics Letters. 2007;**91**:112109

[35] Liu C, Liu W, Chen W, Hsieh S, Tsai T, Yang L. ITO as a diffusion barrier between Si and cu. Journal of the Electrochemical Society. 2005;**152**:G234-G239

[36] Gee JM. Cu Paste Metallization for Silicon Solar Cells. Google Patents; 2011

[37] Li J, Li Y, Wang Z, Bian H, Hou Y, Wang F, Xu G, Liu B, Liu Y. Ultrahigh oxidation resistance and high electrical conductivity in copper-silver powder. Scientific Reports. 2016;**6**:39650

[38] Li J, Mayer J, Colgan E. Oxidation and protection in copper and copper alloy thin films. Journal of Applied Physics. 1991;**70**:2820-2827

[39] Kawamura G, Alvarez S, Stewart IE, Catenacci M, Chen Z, Ha Y-C. Production of oxidation-resistant cu-based nanoparticles by wire explosion. Scientific Reports. 2015;**5**:18333

[40] An BW, Gwak E-J, Kim K, Kim Y-C, Jang J, Kim J-Y, Park J-U. Stretchable, transparent electrodes as wearable heaters using nanotrough networks of metallic glasses with superior mechanical properties and thermal stability. Nano Letters. 2015;**16**:471-478

[41] Kim KO, Kim S. Surface morphology control of cu–Ag alloy thin film on W diffusion barrier by seedless electrodeposition. Journal of Nanoscience and Nanotechnology. 2016;**16**:11701-11706

[42] Rebenklau L, gierth P, Partsch U, Mehlich H, Hausmann J, Grimm M, Stein W, Bell H, Clement C, Vogg F. Low temperature interconnection techniques for high efficiency heterojunction solar cells. In: Proceeding of the 27th European Photovoltaic Solar Energy Conference; 2012

[43] Heng JB, Fu J, Kong B, Chae Y, Wang W, Xie Z, Reddy A, Lam K, Beitel C, Liao C. > 23% high-efficiency tunnel oxide junction bifacial solar cell with electroplated cu gridlines. IEEE Journal of Photovoltaics. 2015;**5**:82-86

[44] Kleider J-P, Alvarez J, Brézard-Oudot A, Gueunier-Farret M-E, Maslova O. Revisiting the theory and usage of junction capacitance: Application to high efficiency amorphous/crystalline silicon heterojunction solar cells. Solar Energy Materials and Solar Cells. 2015;**135**:8-16

[45] Mishima T, Taguchi M, Sakata H, Maruyama E. Development status of high-efficiency HIT solar cells. Solar Energy Materials and Solar Cells. 2011;**95**:18-21

[46] Wang Q, Page M, Iwaniczko E, Xu Y, Roybal L, Bauer R, To B, Yuan H-C, Duda A, Hasoon F. Efficient heterojunction solar cells on p-type crystal silicon wafers. Applied Physics Letters. 2010;**96**:013507

[47] Wood D, Kuzma-Filipek I, Russell R, Duerinckx F, Powell N, Zambova A, Chislea B, Chevalier P, Boulord C, Beucher A. Passivated busbars from screen-printed low-temperature copper paste. Energy Procedia. 2014;**55**:724-732

[48] Wood D, Kuzma-Filipek I, Russell R, Duerinckx F, Powell N, Zambova A, Chislea B, Chevalier P, Boulord C, Beucher A. Non-contacting busbars for advanced cell structures using low temperature copper paste. Energy Procedia. 2015;**67**:101-107

[49] Boulord C, Chevalier PM, Powell NE, Zambova AP. Composition and conductor formed therefrom. Google Patents; 2012

[50] Albaugh JD, Beaucarne GDS, Powell NE, Zambova AP. Photovoltaic cell and method of forming the same. Google Patents; 2012

[51] Beaucarne GDS, Powell NE, Tous L, Wood DA, Zambova AP. Photovoltaic cell and method of forming the same. Google Patents; 2012

[52] Yoshida M, Tokuhisa H, Itoh U, Kamata T, Sumita I, Sekine S. Novel low-temperature-sintering type cu-alloy pastes for silicon solar cells. Energy Procedia. 2012;**21**:66-74

[53] Tokuhisa H, Ise S, Morita S, Tsukamoto S, Tomita M, Yoshida M. Reliability of a printed cu busbar electrode on a conventional silicon solar cell. Japanese Journal of Applied Physics. 2015;**54**:08KD22

[54] Kim DH, Oh SI, Kang SK, Jun BH, Song YA, Kim SJ, Chun BJ. Copper nano paste, method for forming the copper nano paste, and method for forming electrode using the copper nano paste. Google Patents; 2011

[55] Lee YI, Kim DH, Kim JY, Kwon JH, Kim SE. Conductive paste composition for low temperature firing. Google Patents; 2013

[56] Chang W-C, Weng L-W, Chuang C-K, Liang J-X, Yang T-N, Ma W-Y. The preparation of antioxidant copper paste and its application to silicon solar cells. Journal of Nanoscience and Nanotechnology. 2016;**16**:9125-9131

[57] Ma W-Y, Chang W-C, Weng L-W, Liang J-X, Chung C-H, Yang T-N. The preparation of antioxidative copper pastes applicable to HIT solar cells metallization. In: 2016 IEEE 43rd Photovoltaic Specialists Conference (PVSC); 2016; pp. 2949-2951

[58] Sinha A, Sharma B. Preparation of copper powder by glycerol process. Materials Research Bulletin. 2002;**37**:407-416

[59] Chow G, Kurihara L, Kemner K, Schoen P, Elam W, Ervin A, Keller S, Zhang Y, Budnick J, Ambrose T. Structural, morphological, and magnetic study of nanocrystalline cobalt-copper powders synthesized by the polyol process. Journal of Materials Research. 1995;**10**:1546-1554

[60] Kobayashi Y, Inose H, Nakagawa T, Gonda K, Takeda M, Ohuchi N, Kasuya A. Control of shell thickness in silica-coating of au nanoparticles and their X-ray imaging properties. Journal of Colloid and Interface Science. 2011;**358**:329-333

[61] Dong Q, Huang C, Duan G, Zhang F, Yang D. Facile synthesis and electrical performance of silica-coated copper powder for copper electronic pastes on low temperature co-fired ceramic. Materials Letters. 2017;**186**:263-266

[62] Schörner S, Clement C, Bell H, Gierth P, Rebenklau L. Industrial inert drying system for high efficiency heterojunction solar cells. In: 31st European Photovoltaic Solar Energy Conference and Exhibition; 2015; Germany, Hamburg. pp. 887-889

[63] Roncali J. Conjugated poly (thiophenes): Synthesis, functionalization, and applications. Chemical Reviews. 1992;**92**:711-738

[64] Roncali J, Garnier F, Lemaire M, Garreau R. Poly mono-, bi-and trithiophene: Effect of oligomer chain length on the polymer properties. Synthetic Metals. 1986;**15**:323-331

[65] Winter I, Reese C, Hormes J, Heywang G, Jonas F. The thermal ageing of poly (3, 4-ethylenedioxythiophene). An investigation by X-ray absorption and X-ray photoelectron spectroscopy. Chemical Physics. 1995;**194**:207-213

[66] Yabuki A, Arriffin N. Electrical conductivity of copper nanoparticle thin films annealed at low temperature. Thin Solid Films. 2010;**518**:7033-7037

[67] Khanna P, Gaikwad S, Adhyapak P, Singh N, Marimuthu R. Synthesis and characterization of copper nanoparticles. Materials Letters. 2007;**61**:4711-4714

[68] Liu X, Zhou Y. Electrochemical synthesis and room temperature oxidation behavior of cu nanowires. Journal of Materials Research. 2005;**20**:2371-2378

[69] Chen C-H, Yamaguchi T, Sugawara K-i, Koga K. Role of stress in the self-limiting oxidation of copper nanoparticles. The Journal of Physical Chemistry B. 2005;**109**:20669-20672

[70] Foresti E, Fracasso G, Lanzi M, Lesci IG, Paganin L, Zuccheri T, Roveri N. New thiophene monolayer-protected copper nanoparticles: Synthesis and chemical-physical characterization. Journal of Nanomaterials. 2008;**2008**:62

[71] Magdassi S, Grouchko M, Kamyshny A. Copper nanoparticles for printed electronics: Routes towards achieving oxidation stability. Materials. 2010;**3**:4626-4638

[72] Pyo Y, Choi D, Son Y-H, Kang S, Yoon EH, Jung S-B, Kim Y, Lee CS. Fabrication of high quality carbonaceous coating on cu nanoparticle using poly (vinyl pyrrolidone) and its application for oxidation prevention. Japanese Journal of Applied Physics. 2016;**55**:055001

[73] Bokhonov B, Novopashin S. In situ investigation of morphological and phase changes during thermal annealing and oxidation of carbon-encapsulated copper nanoparticles. Journal of Nanoparticle Research. 2010;**12**:2771-2777

[74] Hao C, Xiao F, Cui Z. Preparation and structure of carbon encapsulated copper nanoparticles. Journal of Nanoparticle Research. 2008;**10**:47-51

[75] Jacob DS, Genish I, Klein L, Gedanken A. Carbon-coated core shell structured copper and nickel nanoparticles synthesized in an ionic liquid. The Journal of Physical Chemistry B. 2006;**110**:17711-17714

[76] Li H, Kang W, Xi B, Yan Y, Bi H, Zhu Y, Qian Y. Thermal synthesis of cu@ carbon spherical core–shell structures from carbonaceous matrices containing embedded copper particles. Carbon. 2010;**48**:464-469

[77] Li J, C-y L. Carbon-coated copper nanoparticles: Synthesis, characterization and optical properties. New Journal of Chemistry. 2009;**33**:1474-1477

[78] Gao MR, Xu WH, Luo LB, Zhan YJ, Yu SH. Coaxial metal nano–/microcables with isolating sheath: Synthetic methodology and their application as interconnects. Advanced Materials. 2010;**22**:1977-1981

[79] Wang S, He Y, Liu X, Huang H, Zou J, Song M, Huang B, Liu C. Novel C/cu sheath/core nanostructures synthesized via low-temperature MOCVD. Nanotechnology. 2011;**22**:405704

[80] Wang S, Huang X, He Y, Huang H, Wu Y, Hou L, Liu X, Yang T, Zou J, Huang B. Synthesis, growth mechanism and thermal stability of copper nanoparticles encapsulated by multi-layer graphene. Carbon. 2012;**50**:2119-2125

[81] Lee YI, Lee KJ, Goo YS, Kim NW, Byun Y, Kim JD, Yoo B, Choa YH. Effect of complex agent on characteristics of copper conductive pattern formed by ink-jet printing. Japanese Journal of Applied Physics. 2010;**49**:086501

[82] Eiroma K, Auvinen A, Forsman J, Hult E-L, Jokiniemi J, Koskela P, Sarlin J, Sipiläinen-Malm T, Tapper U. Development of conductive carbon coated copper nanoparticle inkjet fluid. NIP & Digital Fabrication Conference; 2011; pp. 458-461

[83] Dang ZM, Zhang B, Li J, Zha JW, Hu GH. Copper particles/epoxy resin thermosetting conductive adhesive using polyamide resin as curing agent. Journal of Applied Polymer Science. 2012;**126**:815-821

[84] Ohnishi S, Nakasuga A, Nakagawa K. Synthesis of copper particles covered with cobalt-catalyzed carbon nanofibers and their application to air-curable conductive paste. Japanese Journal of Applied Physics. 2017;**56**:07KD03

[85] Gierth P, Rebenklau L, Paproth A, Bell H, Clement C, Vogg F, Sontag D, Mehlich H, Hausmann J, Stein W. Influence of inert curing on polymer paste characteristics on high efficency heterojunction solar cells. In: Proceedings of the 28th European Photovoltaic Solar Energy Conference and Exhibition; Paris. pp. 460-463

[86] Chapiro A. Radiation chemistry of polymers. Radiation Research Supplement. 1964; **4**:179-191

[87] Nakamura K, Takahashi T, Ohshita Y. Novel silver and copper pastes for n-type bi-facial PERT cell. In: 31st European Photovoltaic Solar Energy Conference and Exhibition; 2015; Hamburg, Germany. pp. 536-539

[88] Yoshiba S, Dhamrin M, Yoshida M, Uzum A, Itoh U, Tokuhisa H, Sekine S, Saitoh T, Kamisako K. Cost-effective front contact metalization by copper paste for screen-printed crystalline silicon solar cells. In: Proceedings of the 27th European Photovoltaic Solar Energy Conference and Exhibition; 2012; Frankfurt, Germany. pp. 1730-1732

Chapter 3

Efficient Low-Cost Materials for Solar Energy Applications: Roles of Nanotechnology

Williams S. Ebhota and Tien-Chien Jen

Additional information is available at the end of the chapter

http://dx.doi.org/10.5772/intechopen.79136

Abstract

The generation of energy to meet the increasing global demand should not compromise the environment and the future. Therefore, renewable energies have been identified as potential alternatives to fossil fuels that are associated with CO_2 emissions. Subsequently, photovoltaic (PV) solar system is seen as the most versatile and the largest source of electricity for the future globally. Nanotechnology is a facilitating tool that offers a wide range of resources to resolve material challenges in different application areas. This studies X-rays, energy trilemma, potential nanotechnology-based materials for low-cost PV solar cell fabrication, and atomic layer deposition (ALD). In pursuance of improved performance, PV solar-cell technologies have revolutionized from first-generation PV solar cells to third-generation PV solar cells. The efficiency (19%) of second-generation PV cells is higher than the efficiency (15%) of first-generation cells. The second-generation PV cell technologies include a-Si, CdTe and Cu(In,Ga)Se_2), Cu(In,Ga)Se_2 (CIGS) cells. The third-generation PV cells are organic-inorganic hybrid assemblies, nanostructured semiconductors, and molecular assemblies. This nanocomposite-based technology aims at developing low-cost high efficiency PV solar cells. The nanotechnology manufacturing technique, ALD, is seen as the future technology of PV solar cell production.

Keywords: photovoltaic cell low-cost materials, photovoltaic solar technologies, energy trilemma, CO_2 emission, greenhouse gas emission

1. Introduction

The available energy resources are becoming significantly more interesting due to the transition of the world's energy systems. This transition is orchestrated by the falling technology costs and the improved energy conversion and storage efficiency, coupled with the world stands on greenhouse gas emissions. These trends are expected to continue, with renewables playing a key role. Policymakers and regulatory frameworks of many countries must respond quickly and appropriately to catch up with technology alternatives and unstable energy demands. In a survey, over 50% of energy leaders predicted

Disruptors	**Need trend**
African economic growth	Demand generation
	Growing commercial & industrial sector
	Emerging middle class
	Access to electricity
Shifting the energy mix	New capital
	Wind, solar, biomass
	Gas to power
	Nuclear
	Coal
	Hydro
Changing role of customers	Customer shift
	Consumer to producer
	Self-generation
	Demand managers
	Every company is an energy company
Renewable technology	Renewable technology
	Affordability of new technologies
	Solar PV, storage
	Off-grid solutions
Smart grids, smarter utilities	Smart utilities
	Smarter utility management
	Analytics
	Smart grids
	Smart metering
Changing market structures and dynamics	Market restructure
	Consumer to producer
	Self-generation
	Demand managers
	Every company is an energy company

Table 1. Trends of power disruption in SSA [5].

that in 2025 the share of installed distributed generation capacity will increased by 15% or higher [1]. Basden et al. simply put the estimate as 'a business or a home in North America and Europe goes solar every two minutes' [2]. In 2002, United Nations Development Programme estimated the amount of energy that strikes the earth from the sun per year is 1575–49,837 exajoules (EJ). This is by far more than the world's annual energy consumption of about 559.8 EJ [3]. It was posited in Gratzel study in 2001 that covering 0.1% of the earth's surface with PV panels of 10% efficiency will generate the world's total energy need [4]. And yet the developing countries of Africa and other regions are wallowing in energy poverty leading to high unemployment, abject poverty and terrifying standard of living. Till date the energy generated from the sun is less than 0.1% of the current global energy need.

The energy generated from the sun is limited by certain factors which include cost of producing solar cells and the PV cell converting efficiency. To raise the conversion efficiency of solar materials has attracted a lot of interest. Previous attempts reduce the effects of these factors yielded successes and studies are still on going to break more grounds. The attempts have been multi-criteria improvement approach, which has led to the three categories of PV cell materials today – first, second and third generations of PV materials. However, the world's quest to replaced fossil fuels with alternative energy sources has further stretched study on solar materials and system. Several interventions to alter the present power situation in most developing regions, especially in sub-Saharan Africa and Asia (South India), have not yielded the expected results. The impacts of the interventions have been engulfed by challenges arising from global technologies landscape changes, which led to several disruptions in the industry. These disruptors have caused a paradigm shift in the industry in SSA and other developing countries, and have been classified into six by Deloitte as presented in **Table 1** [6]. Therefore, a fresh and systematic power infrastructure investment will be needed to meet the current and future energy demand in developing countries. Solar energy is seen as the best option for alternative energy source because of its abundance, and environmental friendliness. This chapter aims to analyze the global energy trends in terms of achievements, challenges and outlooks. The study X-rays global energy accessibility and the role of PV solar cell system in achieving global supply of energy with modern energy attributes. Further, the significance of nanotechnology in enhancing the efficiency of PV solar materials was discussed.

2. Global energy challenges

2.1. Access to electricity in developing economies

Developing countries across the regions of the world, especially, sub-Saharan Africa (SSA) and Asia experience a high percentage of inadequate, costly and epileptic power supply

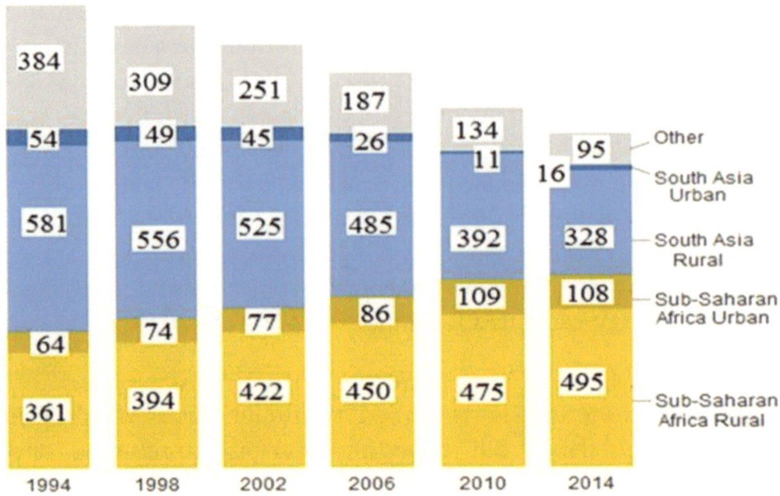

Figure 1. People without electricity access in millions in 2014 [9].

[7, 8], as depicted in **Figure 1**. In 2014, the International Energy Agency (IEA) reported that two-thirds of SSA population has no access to electricity and other modern energy services [10]. Large populations have no access to modern energy in many rural and remote areas of developing countries. It has been predicted that the population without access to electricity in rural areas of SSA would increase from 585 million in 2009 to 645 million in 2030 [11]. The concerted efforts and interventions from both domestic and international arenas to change this scenario have not yielded the expected results.

2.2. Environment sustainability: fossil fuel global environmental challenges

The world's energy demand from fossil fuels accounts for the upward trend in CO_2 emissions, as shown in **Figure 2**. The fossil fuels for power generation are characterized by climate change, greenhouse gases (GHG) emission, and global warming. In 2012, power and transport sectors generated two-thirds of the global CO_2 emission with each emitting about 42% and 23% respectively [13]. For decades, fossil fuels, such as diesel, petrol, coal, and natural gas have proved to be efficient economic development drivers but with health and environmental consequences. Fossil fuels environmental challenges include climate change, global warming, and CO_2 emissions. These negative fallouts have not deterred man from fossil fuel usage because of energy significance to human existence and industrialization. The world's oil consumption annual growth rate of 1.6%, and gas annual growth rate of 1.5% were reported in 2016 [14].

The amount of GHG emission is vastly different amongst countries but is associated with industrialization. The annual CO_2 emissions from fossil fuels combustion have abruptly risen since the Industrial Revolution from near zero to more than 33 $GtCO_2$ in 2015. The highly industrialized countries contribute most to CO_2 emission. The International Energy Agency in 2015 estimated the CO_2 emission from industrial waste and non-renewable municipal waste and the combustion of natural gas, oil, coal and other fuels [15]. **Figure 3** shows 20 of the different countries examined.

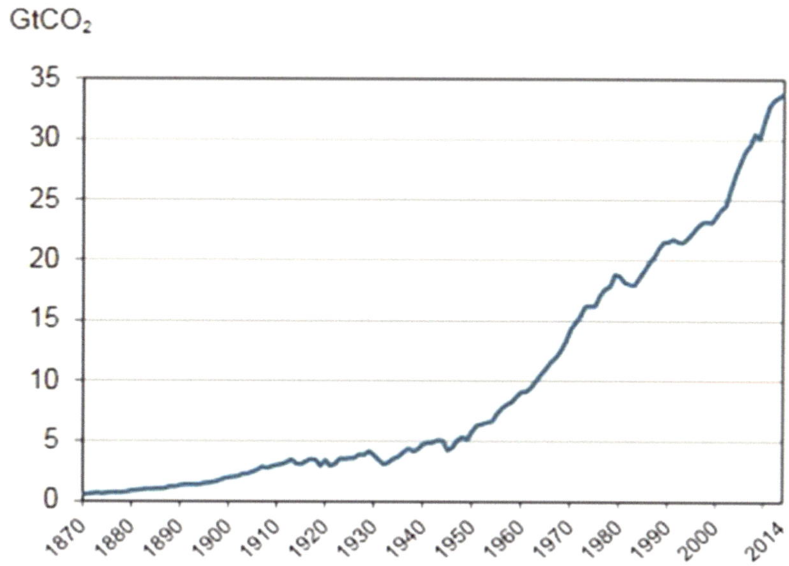

Figure 2. The trend of CO_2 emissions from fossil fuel combustion from 1870 to 2014 [12].

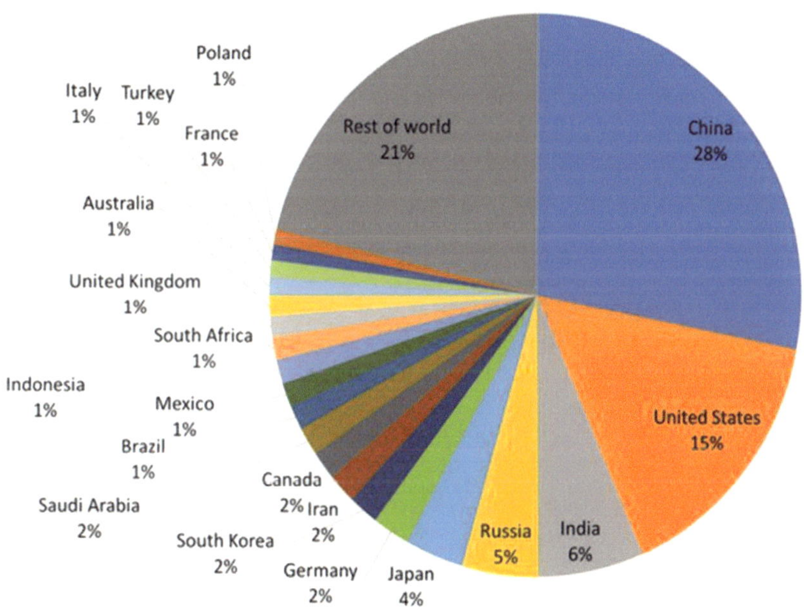

Figure 3. Estimated CO_2 emission of 20 different countries [15].

2.3. Energy trilemma

Apart from the need for adequate, and affordable power supply, the global dynamics in the transformation of electricity sector are these three reinforcing trends–digitization, de-carbonization and decentralization. Also, there is an emergence of authorized energy consumers with new choices in how they utilize and manage their energy usage. The governance and

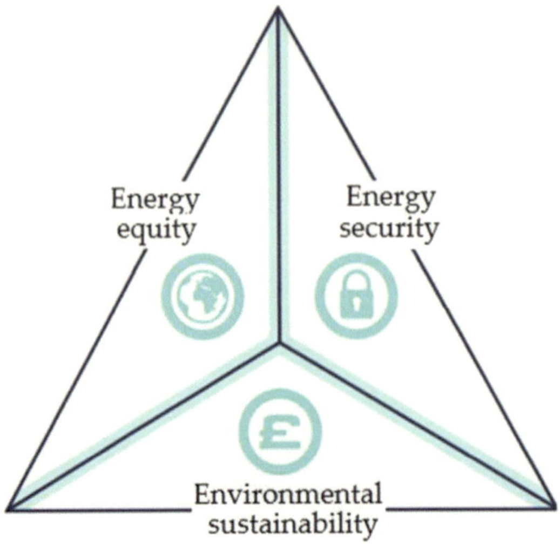

Figure 4. The energy trilemma [16].

management of this complex global energy dynamic is challenging and critical to energy security, climate change mitigation and energy poverty. The interconnectivity between energy security, energy equity and environmental sustainability energy poverty, as shown in **Figure 4**, is termed energy trilemma. Energy sustainability was defined by the World Energy Council based on energy security, energy equity, and environmental sustainability [17–19]. Balancing these three dimensions constitutes what is known as energy trilemma and is the foundation for the individual countries' success and competitiveness [16].

The challenge of balancing energy affordability, energy security, and environmental sustainability could promote the understanding of the framework of the disruptions and opportunities of increased decentralization in the energy system. **Table 2** presents opportunities and challenges associated with trilemma.

Components of trilemma	**Opportunities**	**Challenges**
Energy security	Energy supply shocks due to improved system resilience can be cushioned by diversification and decentralization of energy sources.	Challenges to system management and the possibility of system failure are promoted by the increased system complexity and technology needs.
Energy equity	Rural electrification and energy democratization can be improved by the range of models of energy supply.	Costs increase may be occurred for system and structure establishment to accommodate the increased system complexity and technology needs.
Environmental sustainability	Alterations in the energy mix have the potential to contribute towards environmental degradation reduction, de-carbonization, and reduce lifecycle impacts known with certain energy sources.	Changes to the energy mix have the potential to increase carbon emissions and create life cycle impacts associated with certain energy sources.

Table 2. Trilemma opportunities and challenge.

3. Response to global energy and environmental challenges

To appropriately respond to these global challenges, the need to generate more energy without compromising the future becomes a clarion call. This implies the use of fossil fuel should be limited or eliminated. To effectively do this, clean, reliable and renewable energy sources of energy with low or no GHG emissions must be available. To significantly contribute to the realization of global energy trilemma, both developed and developing countries should incorporate the following into their national infrastructure planning:

- Increasing share of renewables
- Electrification of vehicles and process heat by renewable energy sources
- Robust policy framework and consumer sensitization agenda on energy efficiency to reduce GHG emission, costs and risks
- Strengthening the interdependency and complexity of power systems
- Vibrant consumers' awareness program of new energy alternatives
- Greater accessibility and reliable power supply because of growing dependency on electricity
- The rising threat of cyber-attacks should be met with increased energy infrastructure automation
- The introduction of more disruptive technologies, such as energy storage, PV solar, electric vehicles, and power electronics that might substantially change the energy space as the cost of technology drops.

3.1. Alternative energies-clean energies

Renewable energies have been identified as potential alternatives to fossil fuels. This is due to the notable environment benefits the renewable energies offer, such as reduced CO_2 emission and their off grid utilization. Subsequently, the renewable energy technologies are receiving immense attention. The building of renewable technologies infrastructure to increase the portion of electricity generated from renewable energy has commenced in many countries. Policies and framework have been formulated by different countries to guide the use, the growth and the constitutionality of renewable energy. For example, a bill requiring all the electricity retail sellers to serve 33% of their load with renewable energy by 2020 was signed by the governor of California in 2011 [20].

The available alternatives to the fossil fuels are solar, geothermal, tidal, biofuels, hydro, and wind. A huge capacity of the required electricity can be derived from nuclear energy. However, many countries are skeptical about the use of nuclear energy because of the perceived side effects. They consider the use of nuclear energy for electricity as a risky venture. In Singapore for instance, studies by independent analysts and government agencies have described the existing nuclear plants as too risky for Singapore's small size and dense population [21]. Amongst these natural resources sun (solar), small hydropower, and wind are the most established and are considered better alternatives for environment and cheaper electricity sources in the long term. The exploitability of the solar resource in urbanization is more versatile than other renewable energy sources.

3.1.1. Solar energy: photovoltaic solar cell

Criticisms have trailed renewable energy technology due to their low energy densities, intermittency and region-based resources, making them less suitable for urban applications. Solar is the most common renewable energy whose potential is highly region-dependent. However, the annual direct solar irradiation in some regions exceeds 300 W/m^2. Interestingly, several of the regions that are likely to experience the maximum increase in urbanization are in solar-rich regions. Subsequently, a lot of studies and technical advances have been focused on solar efficiency, structure and cost. This resulted in a drastic drop of solar energy installed price by about 50% since 2010 [22]. Despite this achievement, the efficiency of multi-crystalline silicon photovoltaic cell, which is the widely installed panels, is hovering around 10–17% [23]. However, recent studies have shown PV laboratory efficiency over 40%, using concentrated multijunction cells [24]. This means that the photovoltaics power density could surpass 120 W/m^2 under optimal conditions. For instance, Singapore has an annual average of solar irradiance of 1580 kWh/m^2/year and about 50% more solar radiation than temperate countries. This makes solar photovoltaic (PV) generation as the greatest potential for wider utilization in Singapore. In 2014, Singapore planned to raise solar power from 19 MWp installed capacity to 350 MWp by 2020 and this is 5% of the projected peak electricity demand [25].

The world solar heat collectors' thermal power density average is 67 Wt/m^2 [23, 26] and the use of domestic solar hot-water heaters is on the increase because it is low-cost and compact. Up to 84% of urban households installed solar hot-water heaters on their rooftops [27] and five Australian cities saving approximately 17% energy, using a Trombe wall. Trombe is a technique of collecting and storing of solar thermal energy in the summer for heating in the winter. About 91% of the total energy required in a large residential building in Richmond, VA, is provided from this technology [28].

3.2. Sustainable integrated policies and technologies for urbanization

Sustainable development challenges come with a rise in global urbanization in the dense cities especially in the lower-middle-income countries where the growth of urbanization is rapid. Urban sustainable solutions in the form of integrated policies and technologies are needed globally to lower GHG emissions, reduce the cost of clean energy and guarantee safe energy. The common clean energy challenges in urban are energy intermittency and reliability, cost of installation and low power density. Renewable energies such as wind, hydro and solar have common intermittency and reliability challenges. It is not always windy, sunny and the water level in the source is not always the same. This limits the level of providing a constant power supply to users. There are several approaches that are ongoing in tackling these challenges and these include: the combination of renewable energy sources in a hybrid system; and development of low-cost and efficient renewable energy generation and storage materials.

3.3. Hybrid renewable energy systems

The goal of the combined systems is to cushion the inconsistency supply by a power bolster, either a diesel generator or pumped storage hydroelectric. The bolster runs at low output during low peak hours' demand and increases to full output at peak hours' demands. The combination of renewable energy sources, such as solar, hydro, wind, diesel generator and energy storage units, have been studied extensively in recent years. This combination is often called hybrid renewable energy systems (HRES). Hybrid renewable energy system is a response to challenges of scarce supply of a single renewable resource and intermittent generation challenges. The study on HRES involves modeling, managing and optimizing the different energy systems from design to operation, as shown in **Figure 5** [30–34]. It is vital to consider the system design and operation from a range of time scales because HRES has various stages of life cycle. Irrespective of life cycle (whole life or daily), real-time optimization offers a significant opportunity to handle HRES systematically [29].

The attainment of clean energy based urbanization, large depends on the advances in renewable energy generation technologies, in terms of low cost and efficiency. For a faster, more secure transition from fossil fuels, high efficient and low-cost renewable energy generation and storage materials need to be developed.

3.4. Photovoltaic solar systems

The photovoltaic solar system is composed of different supporting components in addition to PV solar modules. This supporting equipment, often referred to as balance of system (BOS), serves to balance the system and to sustain the operation. The BOS components include controllers, energy storage devices, wiring, grid connections, trackers, mounting hardware, and inverters. However, these components vary from one system to another depending on the scale and application. The different types of PV configurations that work for both Grid-connected and Stand-alone applications are shown in **Figure 6**.

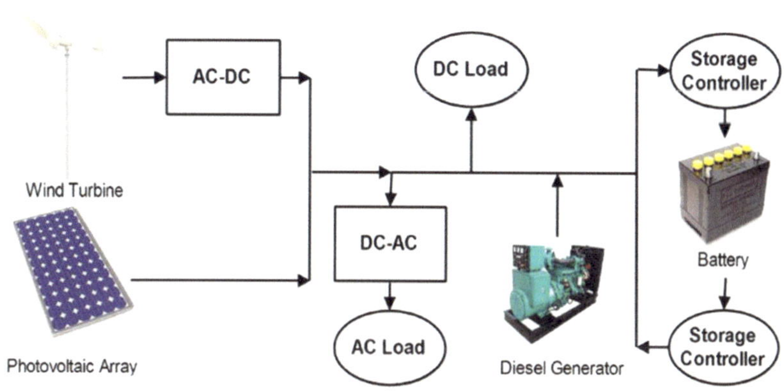

Figure 5. Hybrid renewable energy system (HRES) schematic [29].

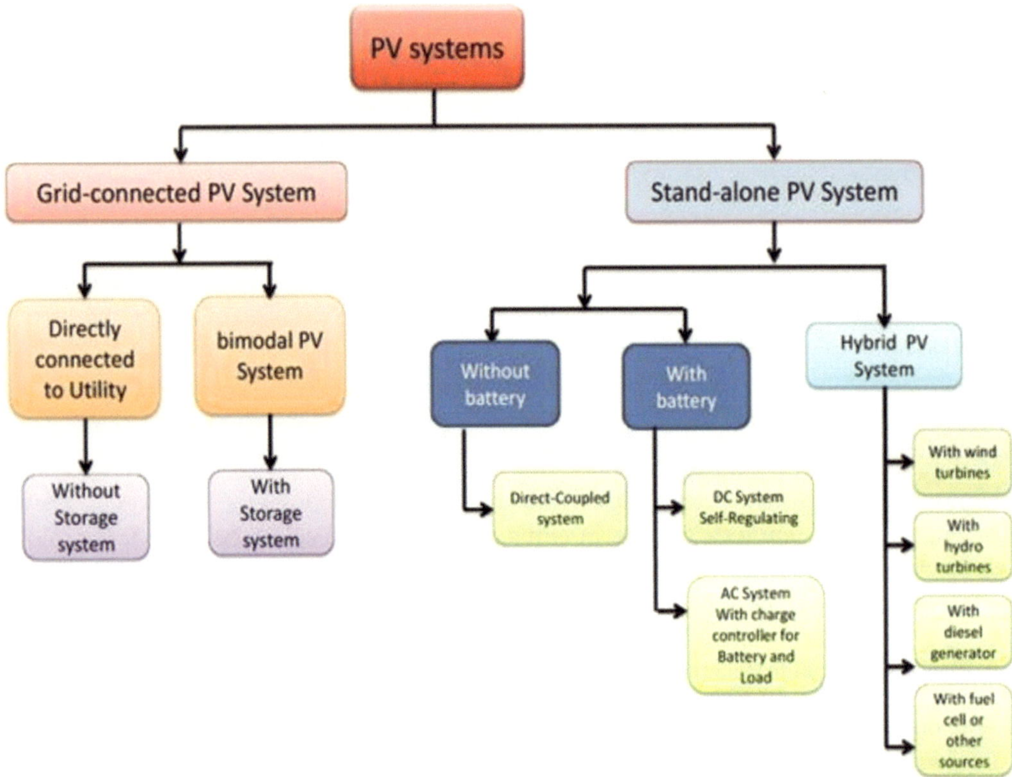

Figure 6. The different types of PV configurations [35].

During repairs, the BOS components can be removed or added to the plant without significant disruption to the infrastructure due to the modular nature of PV solar system. The balance of system components and their functions are presented in **Table 3**. The quality of the BOS is crucial for providing efficient and lasting operation, as the industry aim is to offer PV systems with minimum operational lifetime of 25 years [36, 37]. **Figure 7** shows photovoltaic solar system.

There are three basic processes of photovoltaic effect and they are:

i. Production of charge carriers because of the absorption of photons in the materials that make a junction.

ii. The resulting parting of the photo-generated charge carriers in the junction.

iii. Collection of the photo-generated charge carriers at the terminals of the junction.

The solar cells working principle is based on the production of a potential difference at the junction of two different materials in response to electromagnetic radiation photovoltaic, as shown in **Figure 8**. This occurrence is generally termed photovoltaic effect and is similar to photoelectric effect, which is the emission of electrons from materials that absorb light at a frequency that exceeds the material-dependent threshold frequency. Albert Einstein in 1905 explained this effect has energy and assumed that the light has well-defined energy quanta, called photon, given as Eq. (1):

$$E = hv \quad (1)$$

The absorption of photon by a semiconductor with a bandgap, E_G, is shown in **Figure 8**. The photon with energy $E_{ph} = h_v$ excites an electron from E_i to E_f. At E a hole is created. If $E_{ph} > E_g$ a part of the energy is thermalized.

where h is Planck's constant and v is the light frequency.

Component	Functions
Battery	The electrical power harvested by PV solar panels can be stored in batteries. Solar systems are often built with a battery backup to be used at night or dull days. The batteries are usually deep circle batteries because of their robustness and discharging endurance.
Battery charge controllers	The battery charge controller is similar to automotive battery charger in function. It ensures that a consistent amount of power is sent to the batteries to avoid over charged, and to prevent the backup battery from discharging back through the system at night.
Inverter—solar power converter	It converts the solar panels generated DC power into AC power using electronic switching techniques since home runs on AC power.
Trackers	A device for concentrating a solar reflector, orienting a solar panel or lens towards the sun for optimization of solar energy conversion into electricity.
Mounting hardware	The solar panels are placed on mounting hardware.
Main panel	This is where all electric loads in the building are connected, and protected with circuit breakers.
System power meter	Is enhances the system to gain maximum efficiency from solar installation and estimates the amount of money being saved by the solar system.
Power conditioning unit	It offers protection against electric faults, such as short line-to-ground faults or circuits.
DC and AC disconnect	Solar systems are usually provided with manual disconnection devices for safety reasons. They are used to cut off the power during maintenance and emergency.

Table 3. Functions of photovoltaic solar system components.

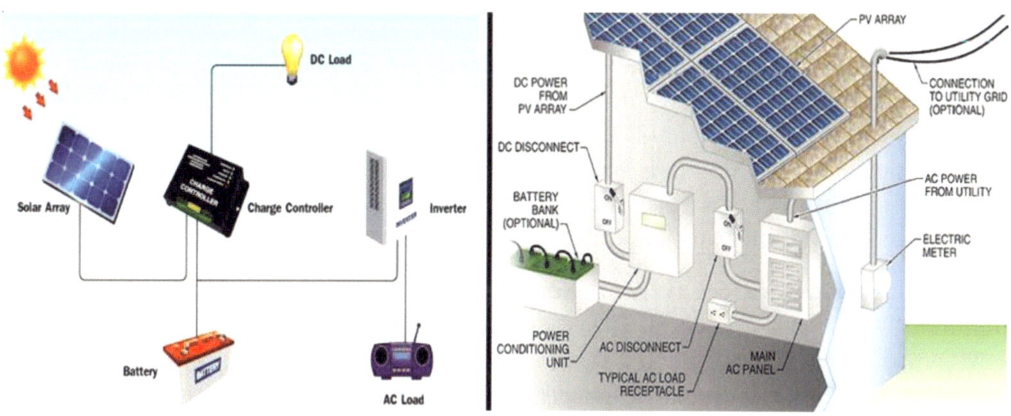

Figure 7. The photovoltaic solar system [38, 39].

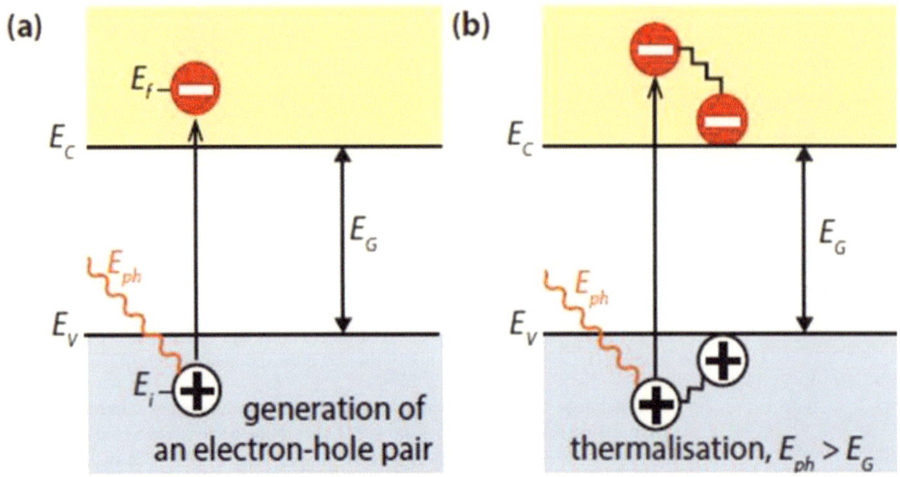

Figure 8. (a) Depicts semiconductor absorption of a photon with bandgap E_G. (b) If $E_{ph} > E_g$ a part of the energy is thermalized.

4. The development of low-cost and efficient renewable energy generation and storage materials

There has been an intense study on the development of low-cost and efficient renewable energy generation and storage materials, but the scope of work only covers PV cell materials. The application of photovoltaic (PV) technology to harness the huge amounts of energy that the sun releases to the earth is one of the most promising alternatives. Thin-film or crystalline silicon are the most widely used materials for industrial production of PV cells [40]. However, the rigidity form of Si PV modules limits their economic incorporation into commercial and residential buildings, while thin film PV cells are more appropriate in terms of cost, ease of fabrication and installation [41]. The thin film PV is receiving more attention and is being considered for large scale power generation and for building integrated photovoltaic (BIPV) applications [42].

4.1. Encapsulation of thin film PV cells

Photovoltaic modules are desirable to provide cheap power for more than 20 years at <10% power degradation outputs at an affordable production cost [43]; and survive 1000 hours at ambient conditions of 85°C and 85% relative humidity in accordance with IEC61646 international standard [44]. Providing thin layer barriers to protect thin film CIGS and DSSC to satisfy these requirements has been challenging. This is because both cells degrade under the ambient conditions if they are not properly protected from moisture ingress.

The requirement for efficient methods to module encapsulation, is a serious challenge facing thin film PV modules producers. The quest to develop appropriate approach for thin film cells encapsulation started around 2002 and is still ongoing. Hence, new low-cost methods of module encapsulation are required to meet this desire. Present developments in the PV cells industry have affected the initial barrier coatings. DSSC and CIGS on flexible substrates are now being considered for flat plate modules with glass-to-glass and BIPV as well as for flat plate module applications, respectively.

4.2. Photovoltaic module materials

Copper indium gallium selenide (CIGS) – one of the most recent developed materials in the renewable energy space, are copper indium gallium selenide, $CuIn_{1-x}Ga_xSe_{2,}$ (CIGS) materials, used for flexible thin-film photovoltaic (PV) modules. Their efficiency levels are beyond that of Si-based rigid PV modules and the films offer substantial advantages in the mass and building integrated photovoltaic (BIPV) applications. However, they are greatly susceptible to environmental degradation in the long term due to water vapor transmission to the active (absorber) layer via the protective encapsulation layer. To prevent the permeability of water vapor to the absorber, a barrier layer of a few nanometres thickness is provided to encapsulate the PV. Roll-to-roll (R2R) atomic layer deposition (ALD) methods are used to deposit amorphous aluminum oxide (Al_2O_3) material on a planarized polyethylene naphthalate (PEN) substrate. Howbeit, water vapor still permeates the barrier because of micro and nano-scale defects present, generated by the deposition process. This occurrence reduces the cell efficiency unit longevity, and ultimately, causes failure [45]. Roll-to-roll technology is used to manufacture flexible devices by repeatedly depositing and patterning of thin layer materials on polymer films substrates [46]. The thickness of the flexible thin layer prior to final encapsulation is about 3 μm [47].

In the solar community, apart from CIGS, dye-sensitized solar cells (DSSCs) have attracted interest and are being given considerable attention due to these attributes: ease of production, comparatively low fabrication cost; and reasonable solar-to-electrical efficiency. Subsequently, they have been regarded as potential materials to replace traditional Si-based solar cells in specialized applications [48]. However, there is a limitation on the absorption coefficient due to standard DSSC employment of a Ru-based dye, which possesses moderate molar absorption [49, 50]. The DSSC system therefore, requires a high surface area substrate for the dye, nanoparticulate TiO_2, and to attain good light harvesting efficiency (LHE). Further, a slow redox shuttle usually I-/I3 is needed since electron transport is reasonably slow in nanoparticulate TiO_2 [51].

5. Application of nano based technology in the manufacturing of energy materials

Chapin et al. at Bell laboratories developed the first generation of PV solar cells, silicon (Si) p/n in 1954 [52]. The first generation PV solar cells have high cost as a major drawback due to the quality of materials used - low defect single Si crystal, reinforced low-iron glass cover sheet, and encapsulants. The second generation solar cells were aimed at lowering cost than the first generation. Examples of the second generation cells technologies are a-Si, CdTe and Cu(In,Ga)Se2), Cu(In,Ga)Se2 (CIGS) cells. Although, this generation cells have efficiency of 19%, which is higher than the first generation PV cells efficiency, the price is considered high. They are usually use in space applications where cost is not a major drawback [53]. In pursuit of overcome the cost and efficiency challenges associated with PV solar cells, the concept of nanocomposite, which enhances device for better performance was utilized. This was exploited in the production of PV solar cells to surmount limits of single materials in solar spectrum response; reaction of holes or electrons with chemicals; transport of holes or electrons; and reduce of costs. This gave birth to the third generation devices that are based on nanocomposites, such as organic–inorganic hybrid assemblies, nanostructured semiconductors, and molecular assemblies. The third generation PV solar cells are aiming to deliver low-cost high efficiency materials nanocomposite.

5.1. Application of atomic layer deposition

Atomic layer deposition (ALD) is a vapor-phase based deposition technique used for depositing high quality, conformal and uniform thin films at comparatively low temperatures. This technique is used to control interface properties through the deposition of high-quality thin films with specific growth control, excellent uniformity over large areas and very good step coverage on non-planar surfaces [54, 55]. These striking attributes of ALD have been employed in several applications including in the fabrication of solar cells for PV modules. In this regard, ALD has benefited the surface passivation layers, buffer layers and barrier layers of crystalline silicon (c-Si), CIGS and dye-sensitized solar cells (DSSCs). Encapsulation of flexible CIGS and organic photovoltaic (OPV) cells with film layer barriers has been performed successfully with ALD. Presently, ALD has been described as the future standard of solar cell equipment manufacturing.

The application of ALD for PV cell started in the 1990s and Bedair and co-workers group reported the first application in 1994. This was followed by the reports on the use ALD to deposit boron-doped ZnO films as transparent conductive oxide (TCO) and ZnSe buffer layers for CIGS solar cells [56]. In 1998, copper indium diselenide (CIS) cells was acclaimed the most common thin film material for PV [57], while the application of a more productive, similar combination material, CIGS, was reported in 2009 [58]. Thin film PV modules are preferred to the conventional crystalline rigid Si cells mainly because of the following merits:

- **i.** Thin film PV cells require less materials [59].
- **ii.** Several vacuum and non-vacuum techniques are used to deposit in thin films PV cells on inexpensive substrates [60].
- **iii.** Thin film deposition on flexible and/or curved substrates, such as polymeric sheets is achievable, forming rollable or foldable solar generators [61]. This has further increased PV cells application areas to include high altitude platforms, cars, aircraft, and various electric appliances. Flexible PV thin-film offers specific design alternatives for BIPVs [62] and the interest in flexible thin film PV cells and technologies is progressively increasing.

In a study [63], ALD was carefully used to develop CuSbS2 thin films at a low-temperature route. After 15 minutes, postprocess, annealing, was performed at 225°C. It was observed that CuSbS2 films ALD-grown, crystalline with micron-sized grains, showed a band gap of 1.6 eV, absorption coefficient of >104 cm^{-1}, and hole concentration of 1015 cm^{-3}. Further, the study demonstrated the first open-circuit voltage on par with CuSbS2/CdS heterojunction PV devices and the potential of ALD grown CuSbS2 thin films in environmentally friendly

http://dx.doi.org/10.5772/intechopen.79136

PVs. It concluded that CuSbS2 and Cu12Sb4S13, members of the Cu–Sb–S family that may be reproducibly synthesized by ALD were added to the short list of metal sulfide thin films.

Recently, attention has been given to new mixed metal oxide materials realized from templating strategy in combination with ALD for DSSC photoelectrodes. Hamann et al. [64] showed the feasibility of the use of TiO_2 structures as dye-sensitized electrodes by typifying their forms and photovoltaic performance. In the study, a new pseudo-one-dimensional structure for DSSC photoanodes was made from mesoporous aerogel, a low-density, high surface area, and thin films by templating. Atomic layer deposition was used to control the variable thickness of TiO_2 deposition on aerogel templates conformally. The study revealed that the cell efficiency was proportional to TiO_2 thickness deposited because of increasing charge diffusion lengths; the electrodes integrated into DSSCs showed good light harvesting and exceptional power efficiencies compared with the nanoparticle TiO_2 based DSSCs; and ALD-coated aerogel-templated photoanodes have been described as a promising candidate to move beyond nanoparticle electrodes in DSSCs due to design flexibility, materials generality, and ease of manufacturing.

6. Conclusion

The World Energy Council categorized energy sustainability into three components - energy security, energy equity, and environmental sustainability. Fossil fuels are the greatest economic drivers amongst the available fuels, but they have huge environmental threats and consequences, such as CO_2 emissions. The increasing global energy demand coupled with the complex nature of the global energy dynamic make management challenging. To effectively respond to the global energy challenges, the generation of energy to satisfy the increasing demand should be done without compromising the environment. Solar energy is a potential alternative to fossil fuel is known for CO_2 emission. As such, photovoltaic solar system is seen as the best option to fossil fuels. The exploitability of solar resource amongst renewable energy sources and the production of low-cost flexible PV cells will facilitate energy trilemma success.

Acknowledgements

The authors would like to thank the financial support from National Research Foundation (NRF) of South Africa and Global Excellence Scholarship from the University of Johannesburg.

Author details

Williams S. Ebhota* and Tien-Chien Jen

*Address all correspondence to: willymoon2001@yahoo.com

Mechanical Engineering Department, University of Johannesburg, Johannesburg, South Africa

References

[1] Austin F. Distributed Disruptions in the Energy Sector. 2017 (Apr 14, 2018). Available: https://www.brinknews.com/distributed-disruptions-in-the-energy-sector/

[2] Basden J, Witkowski A, Wright T. The new make vs buy Calculus how utilities can remain relevant to customers who produce their own power. The Oliver Wyman Energy Journal. 2016;**2**:19-21

[3] Wikipedia Contributors. Solar Energy. In: Wikipedia, The Free Encyclopedia. 2018. Available from: https://en.wikipedia.org/w/index.php?title=Solar_energy&oldid=841654454 [Accessed: May 26, 2018]

[4] Grätzel M. Photoelectrochemical cells. Nature. 2001;**414**:338 (11/15/online)

[5] Ebhota WS, Tabakov PY. Hydropower potentials and effects of poor manufacturing infrastructure on small hydropower development in sub-Saharan Africa. International Journal of Energy Economics and Policy. 2017;**7**:60-67

[6] Deloitte. Sub-Saharan Africa Power Trends: Power Disruption in Africa. (2015, 14/06/2018). Available: https://www.deloitte.com/za/en/pages/energy-and-resources/articles/sub-saharan-africa-power-trends.html

[7] Mohammed YS, Mustafa MW, Bashir N. Hybrid renewable energy systems for off-grid electric power: Review of substantial issues. Renewable and Sustainable Energy Reviews. 2014;**35**:527-539 (Jan 7, 2014)

[8] Trotter PA. Rural electrification, electrification inequality and democratic institutions in sub-Saharan Africa. Energy for Sustainable Development. 2016;**34**:111-129 (Jan 10, 2016)

[9] Khokhar T. Chart: Globally, the Number of People Without Access to Electricity is Falling. 2017. Available from: https://blogs.worldbank.org/opendata/chart-globally-number-people-without-access-electricity-falling [Accessed: Mar 24, 2018]

[10] IEA. 2014 FACTSHEET Energy in Sub-Saharan Africa Today. Paris, France: International Energy Agency; 2014 (World Energy Outlook)

[11] IEA (International Energy Agency). World Energy Outlook 2011. Paris: OECD/IEA; 2011. Available from: https://www.iea.org/publications/freepublications/publication/WEO2011_WEB.pdf [Accessed: Mar 16, 2011]

[12] IEA. CO_2 Emissions From Fuel Combustion Highlights. 2017 ed. France: International Energy Agency (IEA); 2017

[13] IEA. CO_2 Emissions From Fuel Combustion Statistics Highlights. IEA Statistics. 2014 ed. Paris, France: International Energy Agency; 2014

[14] BP. Statistical Review of World Energy. 2017. Available from: https://www.bp.com/content/dam/bp/en/corporate/pdf/energy-economics/statistical-review-2017/bp-statistical-review-of-world-energy-2017-full-report.pdf

[15] IEA. Key Trends in CO_2 Emissions from Fuel Combustion. 2015 ed. France: International Energy Agency (IEA); 2015

[16] Nottingham L, Neher D, Pasquale G, Buser A, Beaven J, Winkler S, et al. World Energy Trilemma: Changing Dynamics – Using Distributed Energy Resources to Meet the Trilemma Challenge. United Kingdom: World Energy Council; 2017

[17] Wyman O. World Energy Trilemma: Priority Actions on Climate Change and How to Balance the Trilemma. London, UK: World Energy Council; 2015

[18] Nentura. Our Energy Trilemma. Apr 2, 2018. Available from: http://www.rdawep.org.au/wp-content/uploads/2017/03/The-Energy-Trilemma.pdf

[19] Tomei J, Gent D. Equity and the Energy Trilemma: Delivering Sustainable Energy Access in Low-Income Communities. London: International Institute for Environment and Development; 2015

[20] EC. California Renewable Energy Overview and Programs. 2011. Available from: http://www.energy.ca.gov/renewables/ [Accessed: Dec 3, 2017]

[21] Zengkun F, Chua G. From The Straits Times Archives: Resolving Singapore's Green Energy Trilemma. 2015. Available from: http://www.straitstimes.com/singapore/environment/from-the-straits-times-archives-resolving-singapores-green-energy-tri-lemma [Accessed: Nov 29, 2017]

[22] U. S. Quadrennial Technology Review: Solar Power Technologies. U.S: U.S. Department of Energy; 2015

[23] Kammen DM, Sunter DA. City-Integrated Renewable Energy for Urban Sustainability. Nov 30, 2017. Available from: https://gspp.berkeley.edu/research/featured/city-integrated-renewable-energy-for-urban-sustainability

[24] NREL. Transforming Energy through Science. USA: National Renewable Energy Laboratory, Office of Energy Efficiency and Renewable Energy; 2016

[25] EMA. Solar Photovoltaic Systems. 2017. Available from: https://www.ema.gov.sg/Solar_Photovoltaic_Systems.aspx

[26] Smil V. Power Density: A Key to Understanding Energy Sources and Uses. USA: MIT Press; 2015

[27] Wei H, Liu J, Yang B. Cost-benefit comparison between domestic solar water heater (DSHW) and building integrated photovoltaic (Bipv) systems for households in urban China. Applied Energy. Jan 8, 2014;**126**:47-55

[28] Terziotti LT, Sweet ML, McLeskey JT Jr. Modeling seasonal solar thermal energy storage in a large urban residential building using TRNSYS 16. Energy and Buildings. Feb 2012;**45**:28-31

[29] Wang X, Palazogluy A, El-Farra NH. Operation of residential hybrid renewable energy systems: integrating forecasting, optimization and demand response. Presented at the American Control Conference (ACC), Portland, Oregon, USA. 2014

[30] Deshmukh M, Deshmukh S. Modeling of hybrid renewable energy systems. Renewable and Sustainable Energy Reviews. 2008;**12**:235-249

[31] Jung J, Villaran M. Optimal planning and design of hybrid renewable energy systems for microgrids. Renewable and Sustainable Energy Reviews. Jan 8, 2017;**75**:180-191

[32] Sawle Y, Gupta SC, Bohre AK. Socio-techno-economic design of hybrid renewable energy system using optimization techniques. Renewable Energy. Nov 21, 2017

[33] Sawle Y, Gupta SC, Bohre AK. Review of hybrid renewable energy systems with comparative analysis of off-grid hybrid system. Renewable and Sustainable Energy Reviews. Jan 1, 2018;**81**:2217-2235

[34] Wang X, Palazoglu A, El-Farra NH. Operational optimization and demand response of hybrid renewable energy systems. Applied Energy. Jan 4, 2015;**143**:324-335

[35] Chaaban MA. PV System Types and Components. Apr 15, 2018. Available: https://www.e-education.psu.edu/ae868/node/872

[36] Baurzhan S, Jenkins G. On-grid solar PV versus diesel electricity generation in sub-Saharan Africa: Economics and GHG emissions. Sustainability. 2017;**9**:372

[37] Ani VA. Design of a stand-alone photovoltaic model for home lightings and clean environment. Frontiers in Energy Research. Jan 11, 2016;**3**: M. A. Chaaban. (15/04/2018). PV System Types and Components. Available: https://www.e-education.psu.edu/ae868/node/872

[38] The Grid Solar. PV Systems: Overview of the Main Components. Apr 10, 2018. Available from: https://thegrid.rexel.com/en-us/energy_efficiency/w/solar_renewable_and_energy_efficiency/82/solar-pv-systems-overview-of-the-main-components

[39] Leonics. Basics of Solar Cell. Apr 10, 2018. Available from: http://www.leonics.com/support/article2_13j/articles2_13j_en.php

[40] Carcia PF, McLean RS, Hegedus S. Encapsulation of Cu(InGa)Se2 solar cell with Al_2O_3 thin-film moisture barrier grown by atomic layer deposition. Solar Energy Materials and Solar Cells. Jan 12, 2010;**94**:2375-2378

[41] Ahmad E. Growth and Characterisation of Cu(In, Ga)Se_2 Thin Films for Solar Cell Applications. UK: University of Salford; 1995

[42] James T, Goodrich A, Woodhouse M, Margolis R, Ong S. Building-integrated photovoltaics (BIPV) in the residential sector: An analysis of installed rooftop system prices. Contract. 2011;**303**:275-3000

[43] Olsen LC. Barrier Coatings for Thin Film Solar Cells Final Subcontract Report. National Renewable Energy Laboratory, Richland, Washington; 2008

[44] CSZ. Photovoltaic Solar Testing Specifications. Nov 16, 2017. Available from: http://www.cszindustrial.com/Products/Custom-Designed-Chambers/Solar-Panel-Testing-Chamber/Solar-Testing-Specifications.aspx

[45] Grover R, Srivastava R, Rana O, Mehta D, Kamalasanan M. New organic thin-film encapsulation for organic light emitting diodes. Journal of Encapsulation and Adsorption Sciences. 2011;**1**:23-28

[46] Galagan Y, Andriessen R. Organic photovoltaics: Technologies and manufacturing. In: Fthenakis V, editor. Third Generation Photovoltaics. Rijeka: InTech; 2012 (Ch. 3)

[47] Elrawemi M. Metrology and characterisation of defects in thin film barrier layers employed in flexible photovoltaic modules. PhD. Mechanical Engineering, University of Huddersfield: UK, 2015

[48] Prasittichai C, Hupp JT. Surface modification of SnO_2 photoelectrodes in dye-sensitized solar cells: Significant improvements in photovoltage via Al_2O_3 atomic layer deposition. The Journal of Physical Chemistry Letters. May 20, 2010;**1**:1611-1615

[49] Martinson ABF, Hamann TW, Pellin MJ, Hupp JT. New architectures for dye-sensitized solar cells. Chemistry – A European Journal. 2008;**14**:4458-4467

[50] Hamann TW, Jensen RA, Martinson ABF, Van Ryswyk H, Hupp JT. Advancing beyond current generation dye-sensitized solar cells. Energy & Environmental Science. 2008;**1**:66-78

[51] Martinson AB, Goes MS, Fabregat-Santiago F, Bisquert J, Pellin MJ, Hupp JT. Electron transport in dye-sensitized solar cells based on ZnO nanotubes: Evidence for highly efficient charge collection and exceptionally rapid dynamics. The Journal of Physical Chemistry A. Apr 23, 2009;**113**:4015-4021

[52] Chapin DM, Fuller CS, Pearson GL. A new silicon P-N junction photocell for converting solar radiation into electrical power. Journal of Applied Physics. 1954;**25**:676-677

[53] Cheng Z. Solar Nanocomposite Materials, Advances in Nanocomposite Technology. London, United Kingdom: IntechOpen; 2011

[54] Puurunen RL. Surface chemistry of atomic layer deposition: A case study for the trimethylaluminum/water process. Journal of Applied Physics. Jun 15, 2005;**97**:121301

[55] George SM. Atomic layer deposition: An overview. Chemical Reviews. Jan 13, 2010; **110**:111-131

[56] Yasutoshi O, Katsumi K, Mitsuru I, Akira Y, Makoto K. Polycrystalline Cu(InGa)Se_2 thin-film solar cells with ZnSe buffer layers. Japanese Journal of Applied Physics. 1995;**34**:5949

[57] Goetzberger A, Knobloch J, Voss B. Crystalline Silicon Solar Cells. Chichester, UK: John Wiley & Sons Ltd; 1998

[58] Inslee J, Hendricks B. Apollo's Fire: Igniting America's Clean Energy Economy. 1st ed. USA: Island Press; 2009

[59] Goetzberger A, Hebling C. Photovoltaic materials, past, present, future. Solar Energy Materials and Solar Cells. 2000;**62**:1-19

[60] Antony A. Preparation and Characterisation of Certain II-VI, I-III-VI2 Semiconductor Thin Films and Transparent Conducting Oxides. Kerala, India: Department of Physics, Cochin University of Science and Technology; 2004

[61] Schubert MB, Werner JH. Flexible solar cells for clothing. Materials Today. Jan 6, 2006;**9**:42-50

[62] Coonen S. Building Integrated Photovoltaics. Nov 15, 2017. Available from: http://ayrintiteknolojileri.com.tr/yuklemedosyalari/file/duyurular/gunesenerjisi/ORNL-Coonen_BIPV.pdf

[63] Riha SC, Koegel AA, Emery JD, Pellin MJ, Martinson ABF. Low-temperature atomic layer deposition of CuSbS2 for thin-film photovoltaics. ACS Applied Materials & Interfaces. Aug 2, 2017;**9**:4667-4673

[64] Hamann TW, Martinson ABF, Elam JW, Pellin MJ, Hupp JT. Atomic Layer Deposition of TiO_2 on Aerogel Templates: New Photoanodes for Dye-Sensitized Solar Cells. The Journal of Physical Chemistry. 2008;**112**:10303-10307

Section 2

Maximum Power Point Tracking for Photovoltaic System

Chapter 4

Improved Performance of a Photovoltaic Panel by MPPT Algorithms

Djamel Eddine Tourqui, Achour Betka,
Atallah Smaili and Tayeb Allaoui

Additional information is available at the end of the chapter

http://dx.doi.org/10.5772/intechopen.79709

Abstract

This work is devoted to the presentation and realization of a digital control card (maximum power point tracking) which serves to improve the performance of a photovoltaic generator (GPV). This makes it possible to increase the profitability of the latter, on the one hand, and the stability of electrical networks, on the other hand. The command card has been developed using simple circuits, and tested on a system that includes a photovoltaic panel powering a resistive load under changing weather conditions. The aim of this paper is to implement three well-known MPPT algorithms (Hill-Climbing, Pertube & Observe and Incremental Conductance), using a PIC microcontroller type 16F877A.

Keywords: photovoltaic panel, MPPT, PIC 16F877A, P&O, Hill-climbing, incremental conductance

1. Introduction

Solar energy is among the most widely used sources of renewable energy on a global scale with an installed global capacity of up to 100 GW [1]. This source is considered one of the most promising and best alternative energy source because of its natural availability and cleanliness [2, 3].

It is known that photovoltaic panels have a non-linear characteristic I = f (V) with a single point where the power generated is maximum (PPM). It is known that PV panels have a non-linear characteristic I = f (V) with a single point where the power generated is maximum (MPP). This maximum power strongly depends on the intensity of solar radiation and the temperature, which changes during the day.

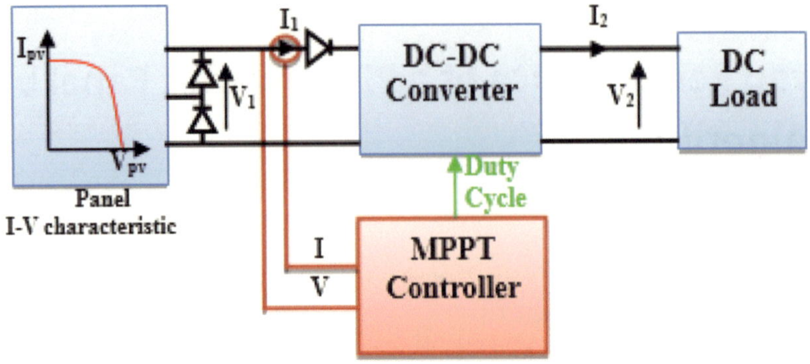

Figure 1. Elementary components of a PV power system.

However, the most difficulties associated with the use of a photovoltaic panel is the perfect non-coupling between the GPV photovoltaic generator and the load [4]. In direct connection mode, a technological barrier that exists in this type of coupling is the problem of transferring the maximum power of the GPV to the load, which often suffers from a bad adaptation. The resulting point of exploitation is then sometimes very far from the real MPP. In other words, it becomes difficult under these conditions to extract the maximum output power of PV panel in all weather conditions [5]. In order to extract at all times the maximum power available at the GPV terminals and transfer it to the load, an MPPT strategy is necessary in order to pursue the maximum power point of the PV panel [6]. There have been many research methods in the literature ranging from the simplest method like Disrupt & Observer (P & O) and IncCond to more sophisticated and complex [7–9].

Static converters, adapted to solar photovoltaic energy, are often called "solar converters" [10]. This adaptation can be achieved by inserting a series chopper controlled by a tracking mechanism "maximum power point tracking" (MPPT). **Figure 1** represents an elementary photovoltaic conversion elementary chain associated with an MPPT control.

2. Modeling of a photovoltaic generator

Figure 2 shows the equivalent electrical circuit (single-diode model) of a photovoltaic generator, which is used to calculate the power supplied by this generator under all irradiation and temperature conditions [11].

The relationship between the cell terminal current I and voltage V is given by [12, 13]:

$$I = I_{ph} - I_D \left[\exp\left(\frac{V + R_s \times I}{\frac{m \times K_B \times T_{amb}}{q}} \right) - 1 \right] - \frac{V + R_s \times I}{R_{sh}} \qquad (1)$$

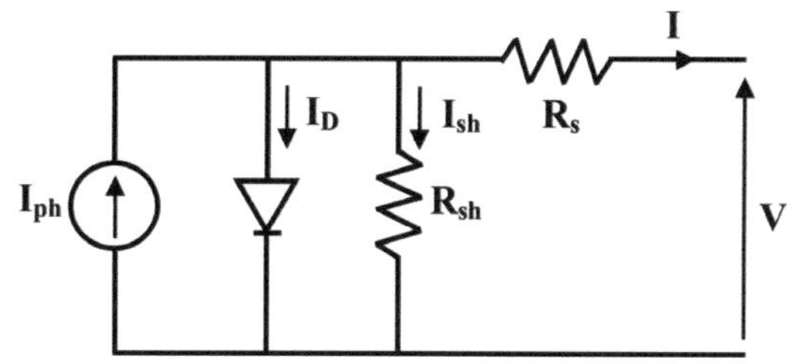

Figure 2. Equivalent circuit of the PV generator.

With:

I: output current, **ID**: reverse saturation diode current, **Iph** photovoltaic current, **KB**: Boltzmann constant (1.3854×10^{-23} JK^{-1}), **m**: ideality factor, **q:** Charge of an electron (1.6021×10^{-19} C), **Rs:** the series resistance (Ω) and **Rsh:** the shunt resistance (Ω).

As mentioned previously, the characteristic I = f (V) (**Figure 1**) of a solar cell strongly depends on the illumination (**E**) and the ambient temperature (**Tamb**) [14]. The empirical model developed by Garcia and Balenzatgui gives the mathematical relation of the temperature of the photovoltaic module as follows [11, 15]:

$$T_m = T_{amb} + \frac{(NOCT - 20)E}{800} \tag{2}$$

In order to calculate the solar generator power (**P**), we used the model developed by Skoplaki and Palyvos [14] as follows:

$$P = E \times A \times \eta_{Tref}\left(1 - B_{ref}\left(T_m - 25\right)\right) \tag{3}$$

3. Overview of MPPT algorithms used

In the literature, there are various examples of MPPT technologies that serve to improve [8, 16, 17]. The Hill-Climbing, IncCond and Perturbe & Observe techniques are the most widely used because of their simplicity and ease of implementation. The operating principle of these three techniques is briefly summarized below:

3.1. Perturb and observe (P&O)

The principle of the P&O type MPPT commands consists in disturbing the panel voltage (V_{PV}) of a small amplitude around its initial value and analyzing the behavior of the instantaneous power variation P_{PV} of the photovoltaic panel before and after the disturbance [16, 18, 19]. If the change in dP_{PV} power increases, this implies that V_{PV} should be set in the same direction as in the previous cycle. If the power of dPPV decreases, it means that the system is far from the optimal point, so the disturbance size must be reduced in order to bring the operating point around to the point of maximum power [20]. In summary, if following a voltage disturbance, the PV power increases, the disturbance direction is maintained. If not, it is reversed to resume convergence to the new MPP. The implementation steps of the P & O technique are illustrated in **Figure 3**.

3.2. Hill-climbing method

The hill-climbing method [16, 21] consists in moving the operating point along the characteristic I = f (V) in the direction in which the instantaneous power PPV increases. For this, the disturbance is applied for the duty cycle D of the converter. The search stops theoretically until the operating power oscillates at the MPP [22, 23]. The flow diagram of this method is illustrated in **Figure 4**.

3.3. Incremental conductance method

To find the MPP, this other technique is based on the knowledge of the GPV conductance variation and the consequences on the position of the operating point with respect to a PPM [24, 25]. Thus, the conductance of the photovoltaic module is defined by the ratio between the current and the voltage of the GPV as indicated below:

The conductance G of the PV circuit is:

$$G = I_{pv} / V_{pv} \tag{4}$$

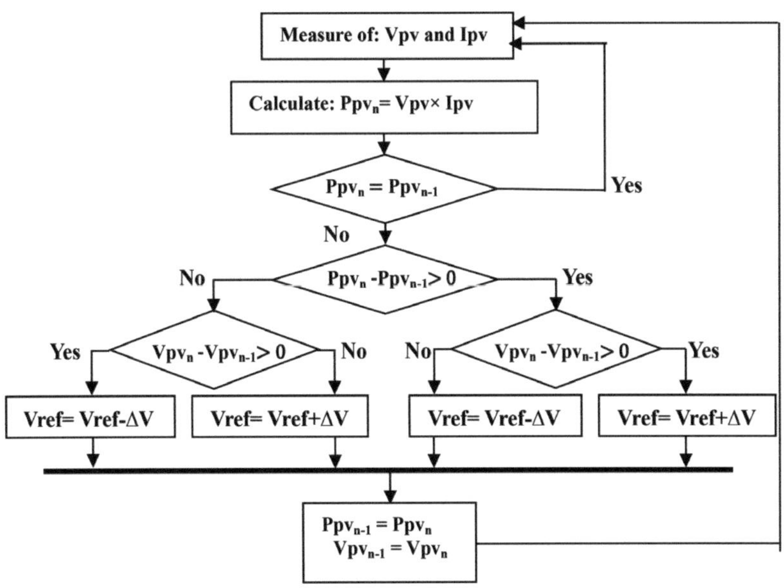

Figure 3. Algorithm of an MPPT command based on the P&O method.

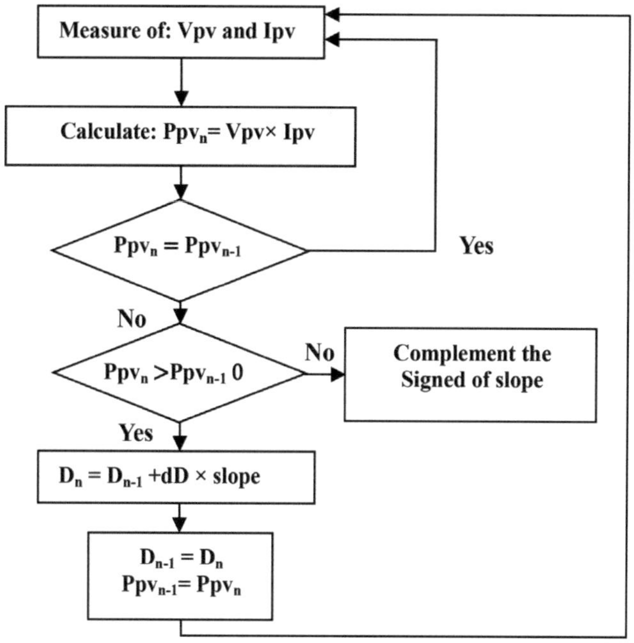

Figure 4. Algorithm of an MPPT command based on the hill-climbing method.

Moreover, an elementary variation (increment) conductance can be defined by:

$$dG = dI_{pv}/\, dV_{pv} \tag{5}$$

Figure 5 shows the position of the operating point on the power characteristic of the PV generator.

The equation of PV panel power is:

$$P_{pv} = V_{pv} \times I_{pv} \tag{6}$$

$$\begin{cases} \dfrac{dP_{pv}}{dV_{pv}} = \dfrac{d\left(V_{pv} \times dI_{pv}\right)}{dV_{pv}} \\ \dfrac{dP_{pv}}{dV_{pv}} = I_{pv} + V_{pv}\dfrac{dI_{pv}}{dV_{pv}} \\ \dfrac{1}{dV_{pv}} \times \dfrac{dP_{pv}}{dV_{pv}} = \dfrac{I_{pv}}{V_{pv}} + \dfrac{dI_{pv}}{dV_{pv}} \end{cases} \tag{7}$$

where:

$$P_{pv} = P_{pvn} - P_{pvn-1},\ dV_{pv} = V_{pvn} - V_{pvn-1} \text{ and } dI_{pv} = I_{pvn} - I_{pvn-1} \tag{8}$$

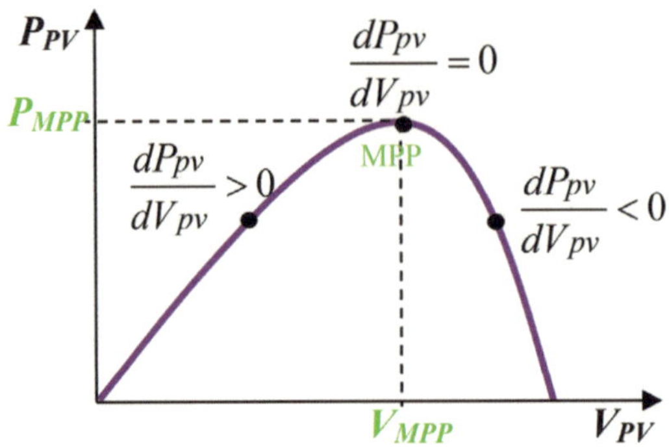

Figure 5. PV power characteristic for different operating points.

On the other hand, the evolution of the power of the module (P_{PV}) with respect to the voltage (V_{PV}) gives the position of the operating point relative to the PPM. When the power derivative is zero, it means that it is on the PPM, if it is positive the operating point is to the left of the maximum, when it is negative, it is to the right of the MPP [23]. **Figure 5** allows to write the following conditions:

$$dP_{pv}/\,dV_{pv} = 0 \quad \text{At the MPP.} \tag{9}$$

$$dP_{pv}/\,dV_{pv} > 0 \tag{10}$$

$$dP_{pv}/\,dV_{pv} < 0 \tag{11}$$

4. Design and realization of the digital MPPT algorithm

At this stage of the research, we will explain the design steps and the realization of the electronic card based on the MPPT algorithms integrated in a microcontroller (μC) PIC. This digital MPPT control based μC has several advantages over analog MPPT control [26, 27]. Our control board contains three important blocks: power block, power supply, and control block.

4.1. Dimensioning of the power block

The control block consists of two essential parts: the measuring circuit is used to read the voltage and current of our photovoltaic panel at the input of the control unit. The second part, which is actually the brain of this block is formed by a microcontroller PIC 16F877A, to program the various proposed MPPT algorithms, and sends the control signal (the duty cycle) of the chopper to the power block, after isolation and amplification.

- **Tensions measurement:** So that the microcontroller can read the voltage of the photovoltaic panel, we must perform the operation of transforming a voltage of 0–22 V into a voltage of 0–5 V.

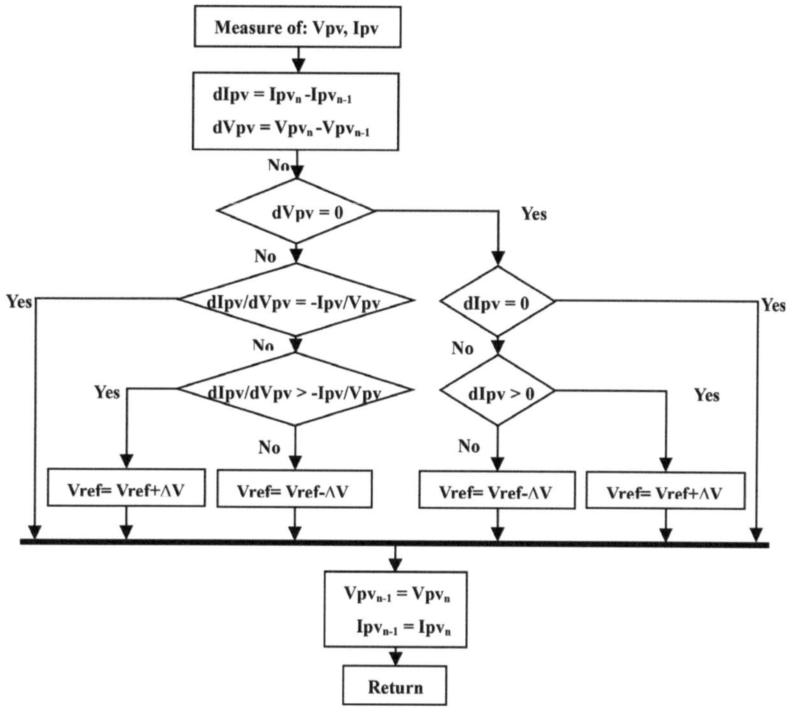

Figure 6. MPPT control algorithm based on IncCond [10].

It is therefore with a simple voltage divider bridge that we perform this operation as show in **Figure 10**. The voltage input to PIC (ΔV_{PIC}) will be connected to pin AN1 of port A configured as input:

Calculation of the resistances:

We choose $V_{PV} = 22(Volts)$ (photovoltaic panel open circuit voltage) and $V_{PIC} = 5(Volt)$ as the maximum input value to the microcontroller:

$$V_{PIC} = \left(\frac{R_2}{R_1 + R_2}\right) \cdot V_{PV} \quad \Rightarrow \quad \frac{V_{PIC}}{V_{PV}} = \left(\frac{R_2}{R_1 + R_2}\right) \tag{12}$$

Digital application: $\frac{5}{22} = \left(\frac{R_2}{R_1+R_2}\right)$.

For: **R_2=1 (KΩ) ⟹ R_1 =3 .4 (KΩ).**

- **Current measurement**

For the measurement of the current derived from the PV module, an inverter amplifier based on an operational amplifier TL082 was chosen. This configuration allowed us to read the value of the current of the panel, with the mass chosen on the side of the load.

The following formulas determine the parameters of this circuit:

$$V_S = \left(-\frac{R_4}{R_3}\right) \cdot V_E \quad \text{With } (V_E = R_{sh} \cdot I_{PV}) \tag{13}$$

So output voltage:

$$V_S = (R_{sh} \cdot I_{PV}) \cdot \left(\frac{R_4}{R_3}\right) \tag{14}$$

With:

$\mathbf{R_{sh} = 0,1(\Omega)}$, V_S: output voltage, V_E: Tension d'entrée input voltage.

4.2. The power block

A Buck converter, or chopper, is a switching power supply that converts a DC voltage into another DC voltage of lower value. Using this converter, the DC input voltage, which is for example generated by the photovoltaic generator (GPV) as shown in **Figure 7**, can be lowered. This serial converter can be used as a source-load adapter, when the direct-coupled operating point is to the left of the MPP. For points to the right of the MPP point, the boost converter is more efficient [28].

It consists of a DC-DC buck converter based on IGBT BUP 314, and ensuring the transfer of all of the power extracted from the solar panel to a resistive load.

If switch **S1** is turned on, diode **D** is reverse biased and a circuit current occurs, but does not pass through this diode (**Figure 8**).

The current iL does not increase immediately, but increases with a rate imposed by inductance **L** [28]:

$$\frac{di_L}{dt} = \frac{V_{PV} - V_{ch}}{L} \tag{15}$$

Meantime, the inductor stores the energy in a magnetic form. If switch **S1** is deactivated after **t = t1**, the load is separated from the source (system supplied). The current is however maintained

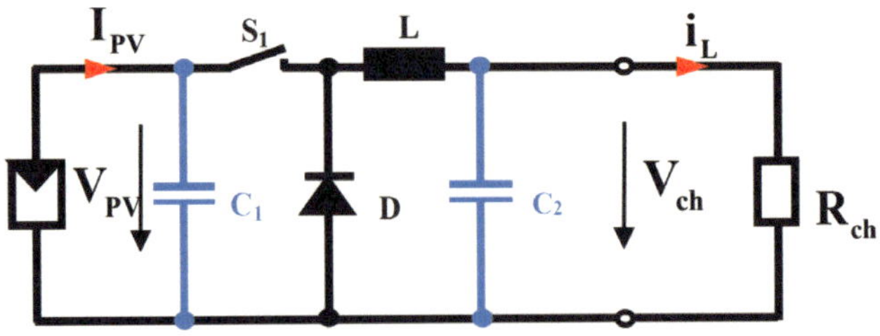

Figure 7. Diagram of the electrical circuit of a Buck converter.

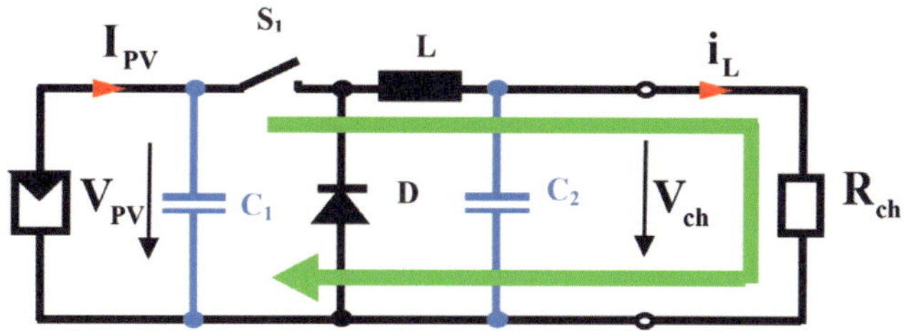

Figure 8. Convertisseur buck Durant l'état on.

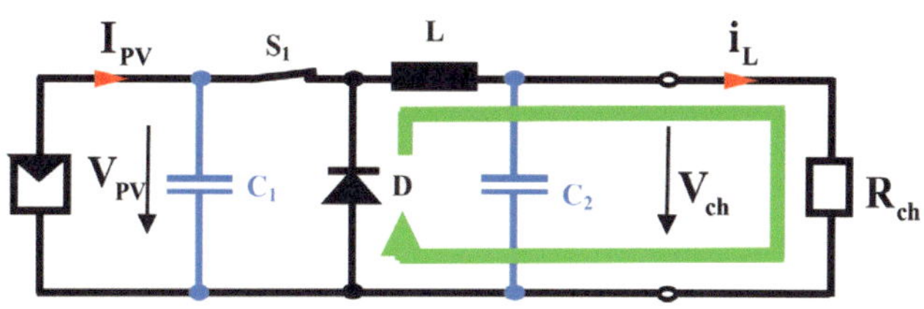

Figure 9. Convertisseur buck Durant l'état off.

Figure 10. PV module employed in the experiment.

by the energy stored in the inductor **L** and flows by means of the freewheeling diode (**Figure 9**). By neglecting the voltage drop across the diode, the current falls, however, because of the following equation:

$$\frac{di_L}{dt} = \frac{-V_{ch}}{L} \tag{16}$$

Capacitor C1 is used to support the supply voltage (**Vpv**). In principle, the switch S1 is activated and deactivated with a switching frequency **f**.

The 80 Watt PV panel used in this study is shown in **Figure 10**.

We operate our serial converter in continuous conduction mode (CCM) and the parameters of this circuit are C1 = 2200 μF, C2 = 200 μF and L = 600 μH. This value of 'L' has been chosen so that the converter operates in TLC according to the following equation [29, 30]:

$$L > \frac{V_{pv}}{4 \times dI_L \times f} \tag{17}$$

4.3. The energy block

The operation of our control circuit requires a power supply at three voltage levels. For this, we realized four power supplies based on a voltage regulator:

- The **LM 7805** voltage regulator to supply the microcontroller with a fixed voltage equal to 5 V.
- The two **LM 7815** and **LM 7915** voltage regulators to provide power required current sensor (−15 V and +15 V), based on an operational amplifier the TL082.
- A second **LM 7815** regulator to power the 4 N25 optocoupler with +15 V voltage. The latter will serve as a driver for the power switch, to ensure the galvanic isolation between the power block and the control block.

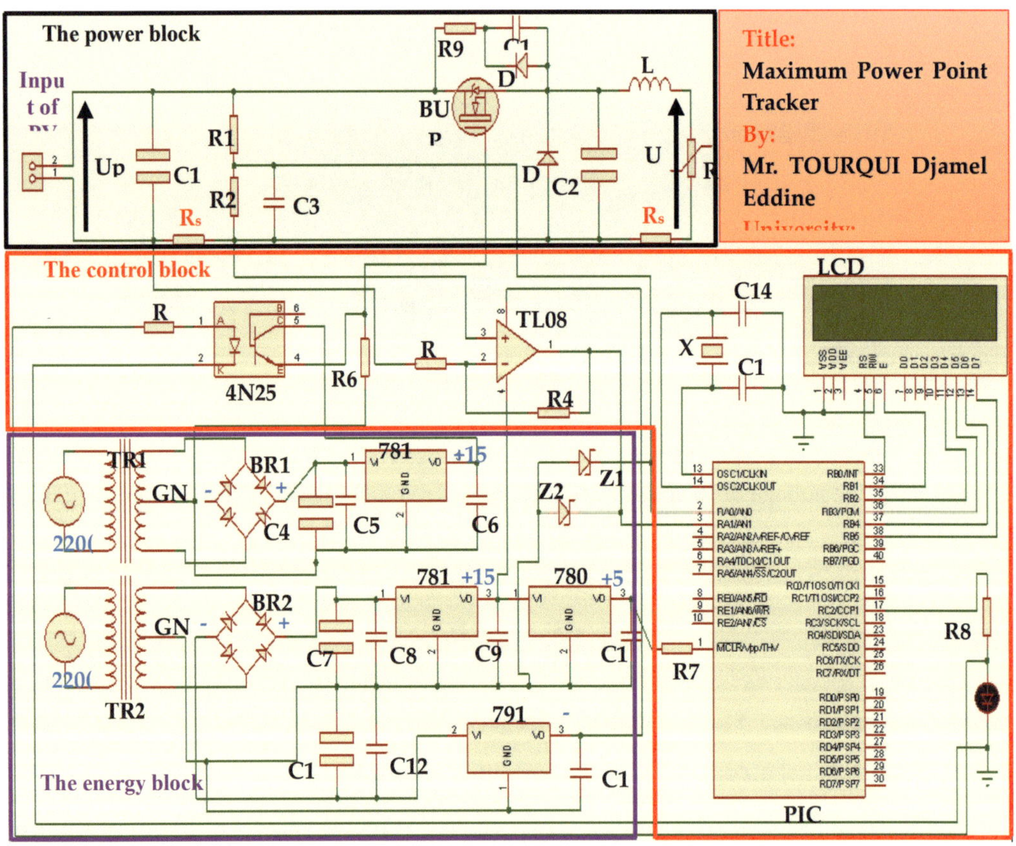

Figure 11. Electrical diagram of the prototype realized.

Figure 12. Practical realization of the electronic cards.

The diagram of **Figure 11**, representing the prototype to be produced, was made under the Proteus PCB design software designed by 'Labcenter Electronics', which makes it possible to draw electronic diagrams, to simulate them and to produce the corresponding printed circuit.

Figure 12a and **b** represents the prototype, which has been realized practically for the digital MPPT control.

5. Results obtained from the MPPT control

In this part, we present the experimental results of the three numerical MPPT algorithms: Perturbed & Observed, Hill-Climbing, and Incremental Conductance, tested on a resistive load (Rm = 3.1 (Ω), which is lower than the load of the maximum power point (MPP) within three (03) clear days. These experiments have been done under the following operating conditions: (1) direct coupling of the load with the photovoltaic panel without MPPT control, (2) using digital MPPT control (DMPPT), (3) by manual MPPT until finding a position to the MPP (manual variation of the load value).

As a result of characteristic I = f(V), we have found that the power generated (P) by the solar panel is related to the intensity of the radiation E and the temperature Tamb. We will take the measurements of: E, Tamb, P and efficiency of control η for each of the three MPPT algorithms studied (Perturb and Observe, Hill-climbing and Incremental Conductance). **Tables 1–3** illustrate the results obtained from the different experiments studied.

The histograms of **Figures 13–15** for these three methods show the difference between the power in the case of direct coupling and the power recovered when applying digital MPPT control that it is compared by the maximum power point search method manually.

Figure 16 illustrates current and voltage (Upv, Ipv) of the photovoltaic generator, current and voltage of the resistive load (Uch, Ich) for direct coupling and MPPT cases based on the P & O technique.

Time (hh:mm)	P for direct coupling (watt)	P for DMPPT (watt)	P in MPP (watt)	η (%)	E (watt/m^2)	Tamb (°C)
10:10	18	43	43	100	672	35.6
11:20	25	54	54	100	809	39.3
15:45	12	40	41	97.5	546	39.5
16:11	9	35	35	100	466	35.5
16:30	6	31	32	96.8	404	34.5

Table 1. Experimental values identified by the P&O control

Time (hh:mm)	P for direct coupling (watt)	P for DMPPT (watt)	P in MPP (watt)	η (%)	E (watt/m^2)	Tamb (°C)
09:58	14	39	42	92.8	547	41.5
10:55	18	48	49	97.9	697	44.3
11:27	22	52	52	100	749	46.5
12:10	22	52	53	98.1	774	45.5
13:50	20	47	47	100	686	45.5

Table 2. Experimental values identified by hill-climbing control

Time (hh:mm)	P for direct coupling (watt)	P for DMPPT (watt)	P in MPP (watt)	η (%)	E (watt/m^2)	Tamb (°C)
13:05	27	54	54	100	774	39.3
13:55	22	53	53	100	835	41
14:25	21.5	54	54	100	772	42
14:46	18	50	50	100	715	45.5
15:12	15	45	45.5	98.9	645	45.5

Table 3. Experimental values identified by IncCond control

The duty cycle of the converter in the case of the P & O algorithm is illustrated in **Figure 17**.

The results of current and voltage of PV panel and the load obtained by Hill-climbing algorithm as shown in **Figure 18**. **Figure 19** explains the duty cycle that controlled the DC-DC converter.

Finally, the same experiment is performed using IncCond control and the results shown in **Figure 20**. **Figure 21** shows the duty cycle generated by the IncCond algorithm.

- **Interpretation and discussion of the results**

The results obtained previously in the power tables and histograms clearly show the efficiency of the electronic control card filled for different control algorithms used. The energy extracted from the solar panel using the digital MPPT technique is very large compared to the direct

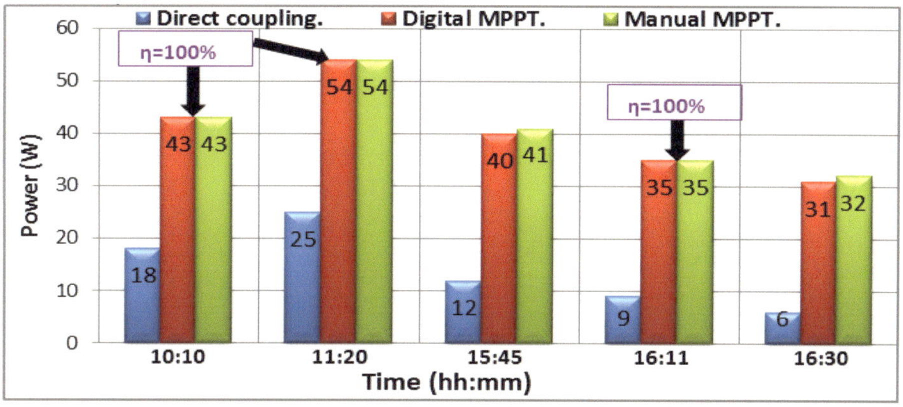

Figure 13. Histogram of powers to P&O algorithm.

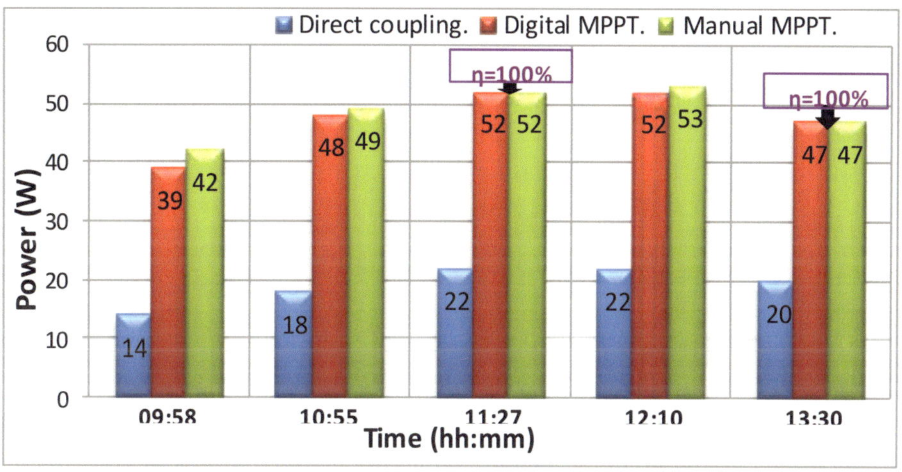

Figure 14. Histogram of powers to hill-climbing algorithm.

charge connection method with the panel. Therefore, in the case of a direct connection between the generator and the load is unlikely to place the PV system at its maximum power point PPM. However, the digital MPPT technique can automatically find the operating voltage of the PV panel that corresponds to the PPM. However, comparing the results obtained by the three algorithms shows the Incremental Conductance technique is the most accurate and closest to the MPP compared to the other two methods.

In addition, the results clearly show the effectiveness of the tracking system (η) which in many cases reaches 100%. This efficiency represents the ratio between the maximum power obtained manually and the other using the MPPT command as indicated in the following equation:

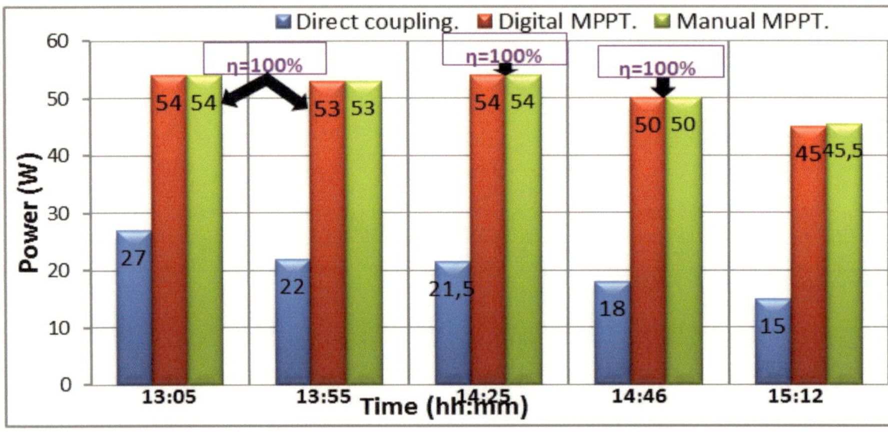

Figure 15. Histogram of powers to incremental conductance algorithm.

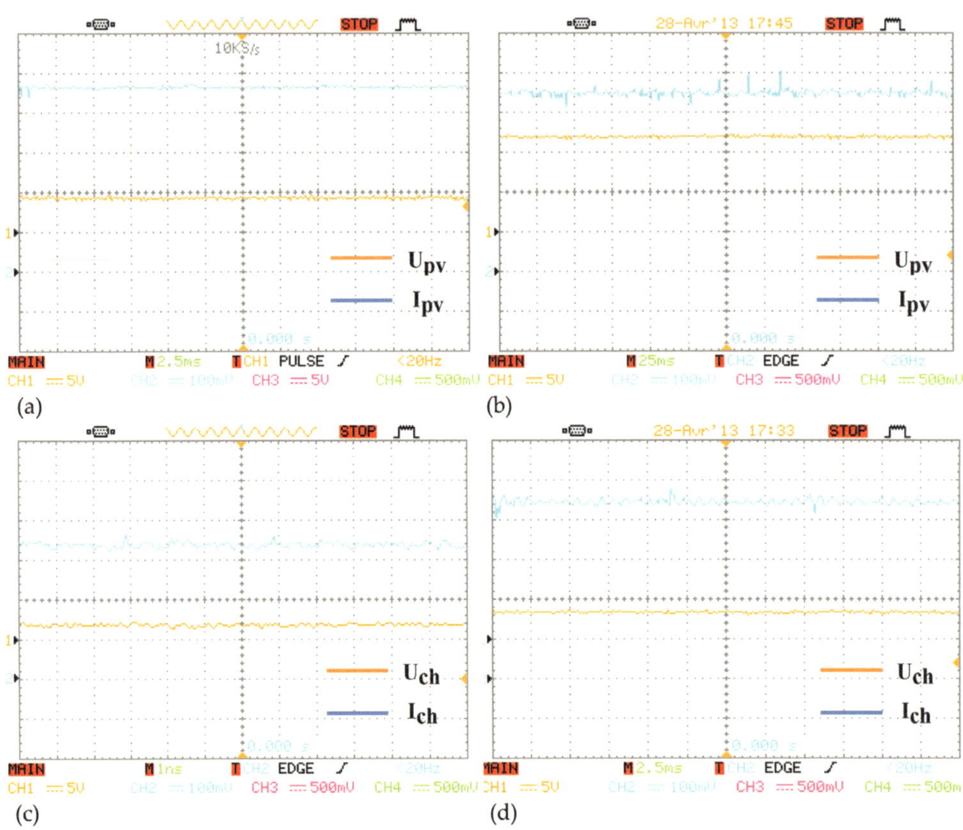

Figure 16. Current and voltage of the photovoltaic generator and the load. (a) (Upv, Ipv) using direct coupling, (b) (Upv, Ipv) using digital MPPT control, (c) (Uch, Ich) using direct coupling, (d) (Uch, Ich) using digital MPPT control.

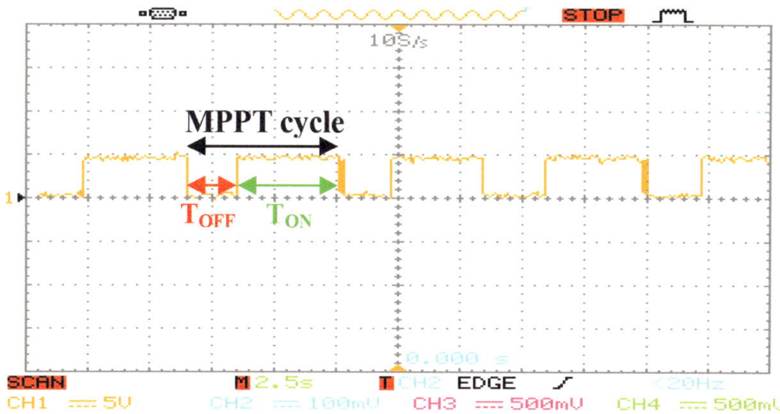

Figure 17. The duty cycle of the P&O algorithm.

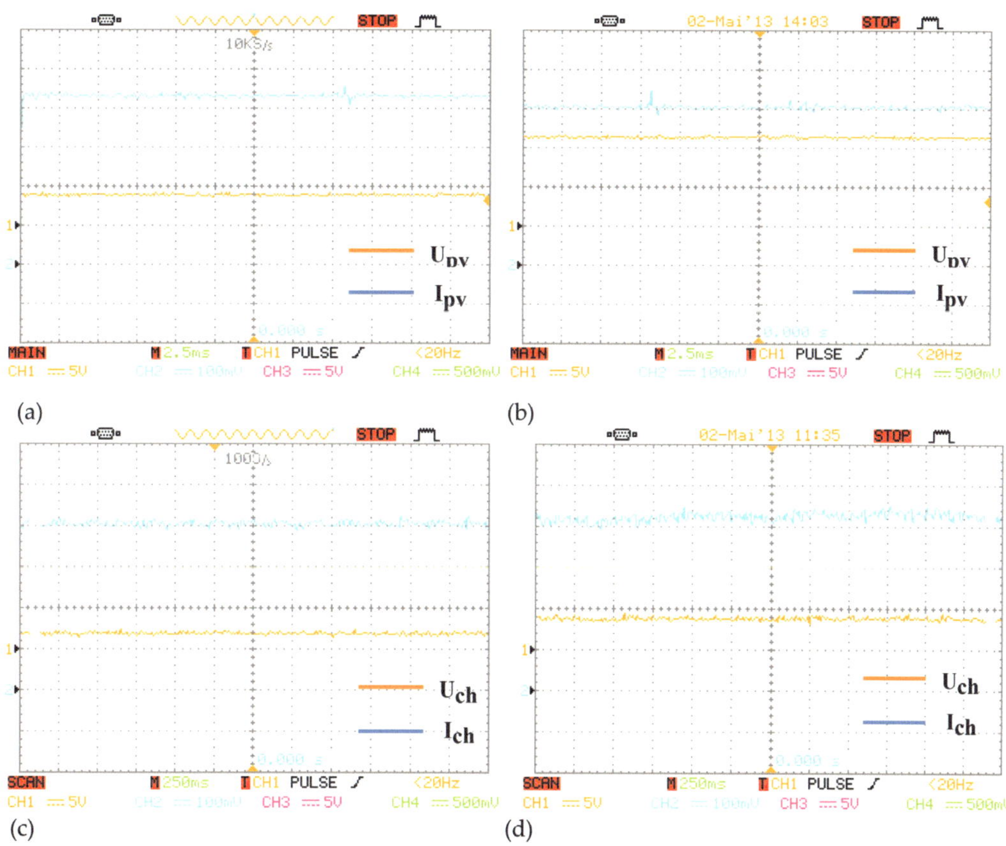

Figure 18. Current and voltage of the photovoltaic generator and the load. (a) (Upv, Ipv) using direct coupling, (b) (Upv, Ipv) using digital MPPT control, (c) (Uch, Ich) using direct coupling, (d) (Uch, Ich) using digital MPPT control.

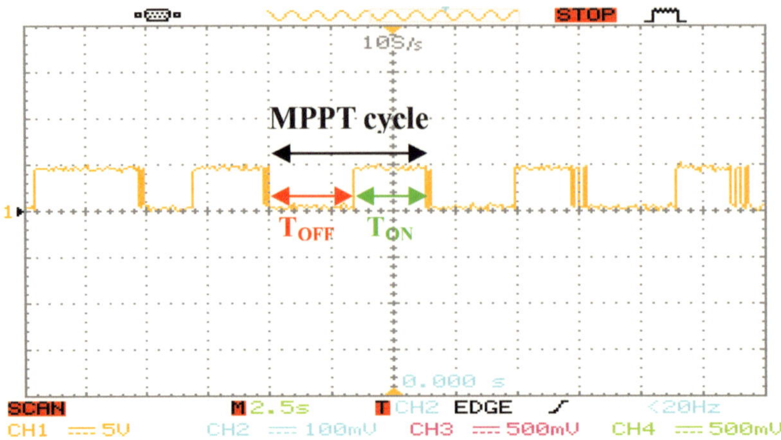

Figure 19. The duty cycle of the hill-climbing algorithm.

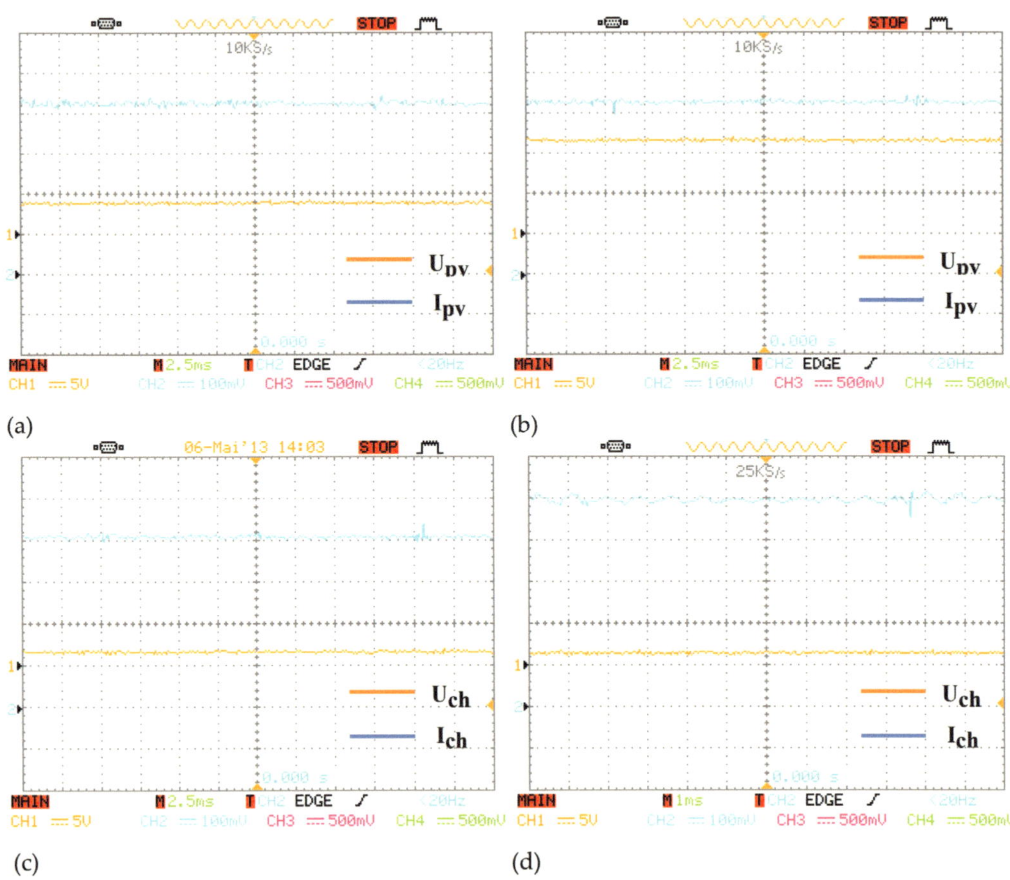

Figure 20. Current and voltage of the photovoltaic generator and the load. (a) (Upv, Ipv) using direct coupling, (b) (Upv, Ipv) using digital MPPT control, (c): (Uch, Ich) using direct coupling, (d) (Uch, Ich) using digital MPPT control.

http://dx.doi.org/10.5772/intechopen.79709

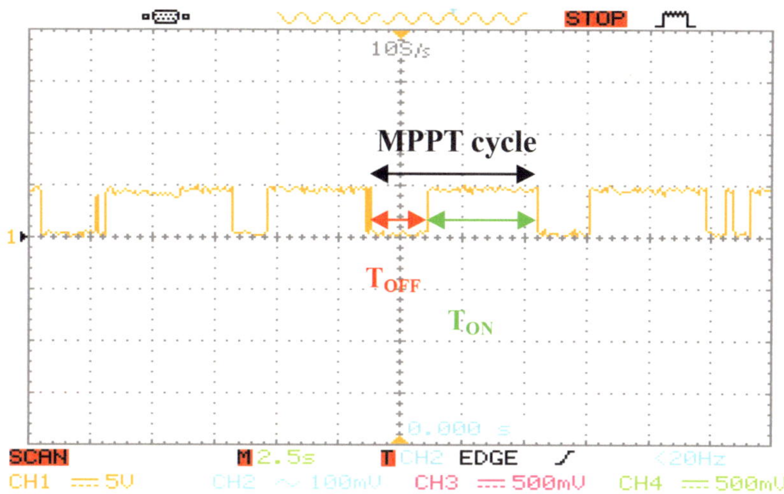

Figure 21. The duty cycle of the incremental conductance algorithm.

$$\eta(\%) = \frac{P_{DMPPT}}{P_{MPP}} \times 100 \tag{18}$$

where P_{DMPPT} represent the power reached by using the proposed of DMPPT controller and $P_{MPP}$is the expected maximum power output in the MPP.

Figures 16, 18 and **20** show that the chopper operates as a voltage step-down, with a voltage of the photovoltaic module stabilizes at Vpv = 14.5 (V). For the current of the load, it is found that the current is in continuous conduction, with a ripple of 2 kHz.

Finally, because of the integration of the PWM control signal into PIC, the duty cycle signal frequency generated by the MPPT command (**Figures 18, 20** and **21**) is of the order of 2 kHz. If the desired maximum power point voltage (VMPP) is higher than the measured panel voltage (Vpv), the duty cycle must be incremented; it must be decreased according to the control technique used. This ratio is adjusted in real time, with the meteorological variations (E and Tamb), and this to position itself on the optimum point.

6. Conclusion and future action

The paper presented a simplified design and implementation of impedance matching stage using a DC-DC buck converter supplying a resistive load controlled by one low cost micro-controller. This circuit allows the acquisition and processing of measured current and voltage signals and generates the appropriate control signals for controlling the switching of the power unit designed primarily around the buck converter. Three popular MPPT algorithms for extracting the maximum power of the photovoltaic panel namely P&O, Hill-Climbing and IncCond have been considered.

MPPT control has led to improved speed of response, a better MPP search accuracy and good control in the presence of perturbations such as sudden variations of the illumination and the temperature.

This work enables us to increase the cost-effectiveness of solar systems as well as reduce the cost which were imported from abroad and the worldwide costly in terms of our scientific laboratory or sector level using this energy in sustainable development agriculture deployed locally as photovoltaic pumping, irrigation and domestic use.

In the future, experiment of these prototypes on other PV installations (like the PV pumping which is available in our laboratory) will be presented in future works.

Author details

Djamel Eddine Tourqui[1]*, Achour Betka[2], Atallah Smaili[3] and Tayeb Allaoui[3]

*Address all correspondence to: tourqui.djamel@gmail.com

1 Morsli Abdellah University Center of Tipaza, Algeria

2 Mohamed Khider University, Biskra, Algeria

3 Ibn Khaldoun University, Tiaret, Algeria

References

[1] Mellit A, Massi PA. Performance prediction of 20kWp grid-connected photovoltaic plant at Trieste (Italy) using artificial neural network. Energy Conversion and Management. 2010;**51**:2431-2441

[2] Parida B, Iniyanb S, Goicc R. A review of solar photovoltaic technologies. Renewable and Sustainable Energy Reviews. 2011;**15**:1625-1636

[3] Mellit A, Kalogirou SA, Hontoria L, Shaari S. Artificial intelligence techniques for sizing photovoltaic systems: A review. Renewable and Sustainable Energy Reviews. 2009;**13**:406-419

[4] Shraif MF. Optimisation et Mesure de Chaîne de Conversion d'Energie Photovoltaïque en Energie Electrique [Phd]. Toulouse, France: Paul Sabatier University; 2002

[5] Orabi M, Hilmy F, Shawky A, Jaber AAQ, Hasaneen E, Gomaa E. On-chip integrated power management MPPT controller utilizing cell-level architecture for PV solar system. Solar Energy. 2015;**117**:10-28

[6] Sivakumar P, Abdullah AK, Yogeshraj K, Arutchelvi M. Analysis and enhancement of PV efficiency with incremental conductance MPPT technique under non-linear loading conditions. Renewable Energy. 2015;**81**:543-550

[7] Anaya-Lara O, Jenkins N, Ekanayake J, Cartwright P, Hughes M. Wind Energy Generation, Modelling and Control. John Wiley & Sons; 2009. ISBN: 978-0-470 71433-1

[8] Esram T, Chapman PL. Comparison of photovoltaic array maximum power point tracking techniques. IEEE Transactions on Energy Conversion. 2007;**22**:439-449

[9] De Brito MAG, Galotto L, Sampaio LP, de Azevedo EMG. Evaluation of the main MPPT techniques for photovoltaic applications. IEEE Transactions on Industrial Electronics. 2013;**60**:1156-1167

[10] Learreta AB. Réalisation de commandes MPPT numériques. Report. Tarragona, Spain: Rovira i Virgili University; 2006

[11] Sharaf Eldin SA, Abd-Elhady MS, Kandil HA. Feasibility of solar tracking systems for PV panels in hot and cold regions. Renewable Energy. 2016;**85**:228-233

[12] Ingegnoli A, Iannopollo A. A maximum power point tracking algorithm for stand-alone photovoltaic systems controlled by low computational power devices. In: 15th IEEE 2010 Mediterranean Electro-Technical Conference; 26–28 April 2010. Valletta, Malta: IEEE. pp. 1522-1527

[13] Villalva MG, Gazoli JR, Ernesto RF. Comprehensive approach to modeling and simulation of photovoltaic arrays. IEEE Transactions on Power Electronics. 2009;**24**:1198-1208

[14] Nema P, Nema RK, Rangnekar S. A current and future state of art development of hybrid energy system using wind and PV-solar. Renewable and Sustainable Energy Reviews. 2009;**13**:2096-2103

[15] Garcia A, Balenzategui JL. Estimation of photovoltaic module yearly temperature and performance based on nominal operation cell temperature. Renewable Energy. 2004;**29**: 1997-2010

[16] Cabal C. Optimisation énergétique de l'étage d'adaptation électronique dédié à la conversion photovoltaïque [Phd]. Toulouse, France: Paul Sabatier III University; 2008

[17] Hohm DP, Ropp ME. Comparative study of maximum power point tracking algorithms using an experimental programmable maximum power point tracking test bed. In: IEEE 2000 Photovoltaic Specialists Conference; 15-22 September 2000; Anchorage, Alaska, USA: IEEE. pp. 1699-1702

[18] Femia N, Petrone G, Spagnuolo G, Vitelli M. Optimization of perturb and observe maximum power point tracking method. IEEE Transactions on Power Electronics. 2005;**20**:963-973

[19] Sera D, Kerekes T, Teodorescu R, Blaabjerg F. Improved MPPT algorithms for rapidly changing environmental conditions. In: Power Electronics and Motion Control Conference 2006; 30 August-1 September 2006; Portoroz, SLOVENIA: IEEE. 2006. pp. 1614-1619

[20] Onat N. Recent developments in maximum power point tracking technologies for photovoltaic systems. International Journal of Photoenergy. 2010;**2010**:11. Article ID 245316. DOI: 10.1155/2010/245316

[21] Xiao W, Dunford WG. Evaluating maximum power point tracking performance by using artificial lights. In: IEEE 2004 Industrial Electronics Society; 2–6 November 2004. Busan, Korea: IEEE; 2004. pp. 2883-2887

[22] Shimizu TH, Kimura OG. A novel high performance utility interactive photovoltaic inverter system. IEEE Transactions on Industrial Electronics. 2003;**18**:704-711

[23] Nur AK, Chee WT. A comprehensive review of maximum power point tracking algorithms for photovoltaic system. Renewable and Sustainable Energy Reviews. 2014;**37**:585-598

[24] Lee JH, Bae H, Cho BH. Advanced incremental conductance MPPT algorithm with a variable step size. In: Power Electronics and Motion Control Conference; 30 August-1 September 2006, 2006. Portoroz, SLOVENIA: IEEE. pp. 603-607

[25] Kim TY, Ahn HG, Park SK, Lee YK. A novel maximum power point tracking control for photovoltaic power system under rapidly changing solar radiation. In: IEEE 2001 International Symposium on Industrial Electronics; 12–16 June 2001. Busan, Korea: IEEE; 2001. pp. 1011-1014

[26] Oi A. Design and simulation of photovoltaic water pumping system [MSc]. California, USA: California Polytechnic State University; 2005

[27] Pongratananukul N. Analysis and simulation tools for solar array power systems [Phd]. Florida, USA: Central Florida Orlando University; 2005

[28] Luque A, Hegedus S. Handbook of Photovoltaic Science and Engineering. USA: John Wiley & Sons Ltd; 2003

[29] Group 01gr509. Power Supply for the AAU Cubesat. Report. AALBORG University; 2001

[30] Erickson RW. Fundamentals of Power Electronics. New York, NY: Chapman & Hall; 1997

Chapter 5

A Quick Maximum Power Point Tracking Method Using an Embedded Learning Algorithm for Photovoltaics on Roads

Koichiro Yamauchi

Additional information is available at the end of the chapter

http://dx.doi.org/10.5772/intechopen.79711

Abstract

This chapter presents a new approach to realize quick maximum power point tracking (MPPT) for photovoltaics (PVs) bedded on roads. The MPPT device for the road photovoltaics needs to support quick response to the shadow flickers caused by moving objects. Our proposed MPPT device is a microconverter connected to a short PV string. For real-world usage, several sets of PV string connected to the proposed microconverter will be connected in parallel. Each converter uses an embedded learning algorithm inspired by the insect brain to learn the MPPs of a single PV string. Therefore, the MPPT device tracks MPP via the perturbation and observation method in normal circumstances and the learning machine learns the relationships between the acquired MPP and the temperature and magnitude of the Sun irradiation. Consequently, if the magnitude of the Sun beam incident on the PV panel changes quickly, the learning machine yields the predicted MPP to control a chopper circuit. The simulation results suggested that the proposed MPPT method can realize quick MPPT.

Keywords: photovoltaics bedded on road, embedded learning algorithm, incremental learning, insect brain, modal regression on a fixed memory budget, maximum power point tracking (MPPT), shadow flicker, partial shading, micro converter

1. Introduction

In recent years, renewal energy technologies have attracted considerable attention as they prevent degradation of the environment to a large extent. Photovoltaics (PVs) are one such technology. However, the drawbacks of photovoltaic systems are that they are unstable while generating electricity and that they require a wide area to catch a large amount of sunlight.

One solution is to place photovoltaics on roads. As the total area covered by roadways in the world is extremely high, it is worth using it as PV sites. Still, objects moving on the road cause shadows. In particular, the shadow flickers on PV systems cause power conditioners connected to the PVs to behave in an unstable manner. Such unstable behavior forms the origin of degradation and greatly reduces the amount of electricity generated.

As shown in Section 2, PVs demonstrate highly nonlinear characteristics and its maximum power point cannot be analytically derived. Therefore, maximum power point tracking (MPPT) devices track MPP using various heuristics. As mentioned in previous survey papers [1, 2], the most preliminary technique for realizing MPPT is the perturbation and observation (P&O) method. P&O is a type of hill-climbing algorithm. The P&O method provides a perturbation to the current and the voltage and checks whether the output power increases. If the power has increased, the P&O method employs the same voltage change in the next step and vice versa. Although the P&O method is easy to implement within small embedded systems, there is no guarantee that the perturbed voltage is suitable for obtaining MPP. The incremental conductance (IncCond) [3] and the ripple correlation (RCC) methods [4] overcome this problem by estimating the gradient of the power curve. These two methods can be realized in analog circuits and can demonstrate quick convergence behaviors. Fuzzy logic control methods are also usually used for controlling the change in duty ratio for the chopper circuit. Fuzzy logic controllers can work appropriately even if its inputs are ambiguous, and they show a quick convergence behavior to the MPP. For example, a previous paper [5] demonstrated the use of fuzzy logic that yielded a change in the duty ratio from the difference between the current photovoltaic output voltage and the predicted MPP. Neural network-based MPPT methods are also proposed (e.g., [6]). The model predicts MPP and its corresponding maximum current using a pretrained neural network. The model cannot adjust its neural network for changing environments. In our previous study, a hybrid system involving the P&O method and an embedded learning machine was constructed [7]. The learning machine studies the MPP acquired by the P&O method when solar irradiation is stable. When solar irradiation changes quickly, the learning machine predicts MPP. However, these methods do not support MPPT under an inhomogeneous isolation condition, where the voltage-power curve has several local peaks.

Recently, a particle swarm optimization (PSO)-based MPPT method was proposed [8]. This method can estimate all local power peak points and select the best one. However, the resultant solutions are highly depending on the initial particles.

On the contrary, a previous study [9] demonstrated that a swing technique can acquire the voltage-power curve by scanning within a certain short interval. It shorts, the series-connected PV string and an inductor simultaneously observe the voltage and power until the output voltage reaches zero. Therefore, the device can detect MPP during the scan. However, it needs special hardware to realize the swing.

To overcome this problem, we use a quick converter connected to a PV string. The main challenge here is finding the MPP from the complex power-voltage curve.

In our previous study [7], we proposed a model that uses an incremental learning method based on general regression neural network. The method is used to obtain the magnitude of solar irradiation s_t, temperature T_t, and MPP derived by the P&O method. Although the system quickly detects MPPs of a single solar panel, which has a single cluster, it cannot detect the MPP of solar panels with several clusters or solar panels connected in series.

In this chapter, we propose an MPPT converter that detects MPPs of solar panels with several clusters using a modal regression method on a fixed memory budget. To realize quick MPPT, the proposed method uses a learning machine on a fixed memory budget. The learning machine on a fixed budget is a small learning machine that can continue online learning on a fixed storage space. Therefore, it is suitable to be embedded to a small microcomputer. The learning on a budget should be executed on a system with a small amount of storage space with low computational power.

To this end, it is worth referencing the mechanisms of an insect's brain. Although the precise mechanism of an insect's small brain that is a source of their intelligence is not known, it is true that their sensory system is much smaller than that of humans. Therefore, the dimensions of their sensory inputs are small. As mentioned in Section 3.2, the storage space for recording the kernels is proportional to the number of input dimensions. From this insight, we should be able to reduce the input dimensions to reduce the storage space for the learning machine.

The rest of the chapter is organized as follows. Section 2 describes the photovoltaic properties, and Section 3 introduces an MPPT algorithm accelerated by a learning machine using a modal regression on a budget. Section 4 shows computer simulation results of the new MPPT algorithm, and Section 5 concludes this chapter.

2. Properties of photovoltaics

Photovoltaics are a type of current sources, whose current flow is determined by the strength of solar irradiation. A normal solar panel comprises several photovoltaic cells. These cells are usually connected in series, and the series-connected cells are then connected in parallel. Such solar panels show highly nonlinear characteristics and is usually modeled by using the following equation [10, 11]. Let us denote the output voltage and current from the photovoltaic as V_{pv} and I_{pv}, respectively. According to the equivalent circuit shown in **Figure 1**, I_{pv} is represented by (1).

$$\mathrm{I_{pv}} = \mathrm{N_p} I_{sc}\left(\frac{I_r}{100}\right) - \mathrm{N_p} I_o\left[\exp\left(\frac{qV_{pv}}{nkTN_s}\right) - 1\right], \tag{1}$$

where V_{pv}, the terminal voltage of the photovoltaic [V]; I_{pv}, output current from the photovoltaic [A]; I_p, photocurrent [A]; I_o, saturation current [A]; I_{sc}, short-circuit current [A]; I_r, irradiation [%]; n, ideality factor; q, charge of electron [C]; k, Boltzmann's constant; T, junction temperature [$^\circ$C]; $\mathrm{N_p}$, number of cells in parallel; $\mathrm{N_s}$, number of cells in series.

In Eq. (1), I_r is given by the ratio of actual strength of solar irradiation to the irradiation of standard test condition [11]. Therefore, $I_r = 100G/G_{ref}$, where G and G_{ref} are solar irradiation (w/m^2) and that of under the standard test condition: $G_{ref} = 1000(w/m^2)$, respectively. The range of I_r is $I_r \in [0, 100]$. An example of the output voltage and current relationship is shown in **Figure 2**. We can see that the solar panel is a type of current sources, but the current is

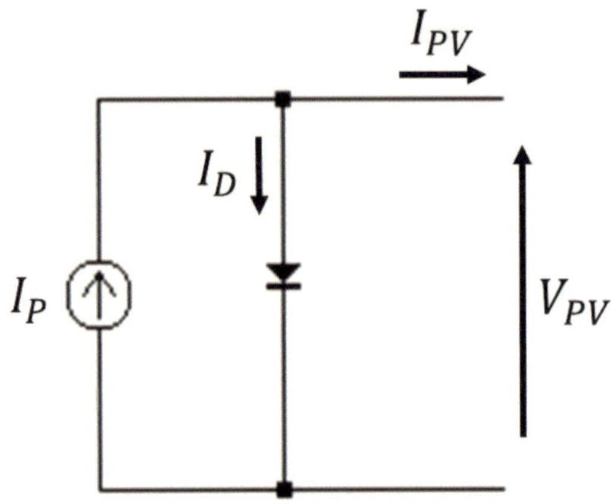

Figure 1. Equivalent circuit of a photovoltaic.

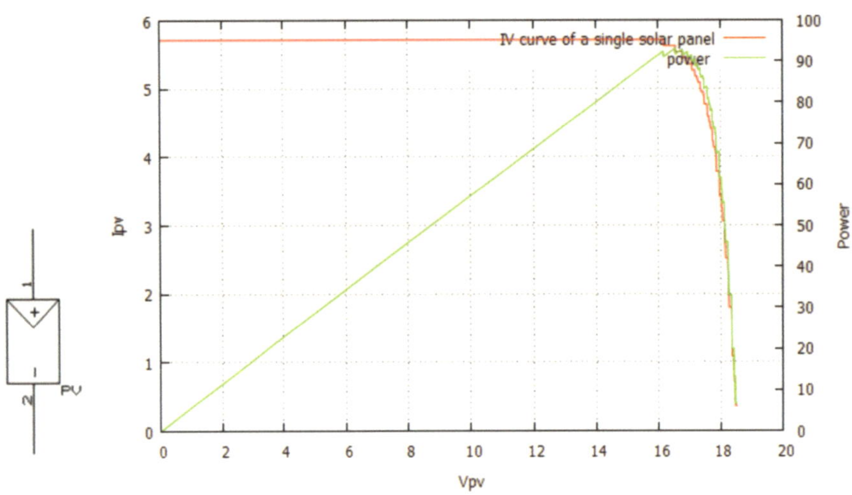

Figure 2. Single solar panel property. (Ir = 80%, $N_p = 4, N_s = 12, I_{sc} = 1.8[A], T = 298.15, q = 1.6 \times 10^{-19}$).

reduced when the voltage is higher than a certain value. The solar panel does not pass current if the panel is covered by a shadow. If series-connected solar panels have a partial shadow, the output current from the solar panels are down to zero even if a part of solar panels do not have a shadow. To prevent such a situation, a bypass diode is connected to each solar panel in parallel. Using this circuit, the solar panels can generate a certain amount of electricity even if they are partially shadowed. Such series-connected solar panels, however, show highly nonlinear characteristics (see **Figure 3**).

To extract maximum power, the voltage of the photovoltaic should be maximized. However, if the voltage is too high, the current decreases. Therefore, there is an optimal voltage value that maximizes the power. Such voltage is called the MPP and the power conditioner or converter connected to the PV tracks the MPP.

http://dx.doi.org/10.5772/intechopen.79711

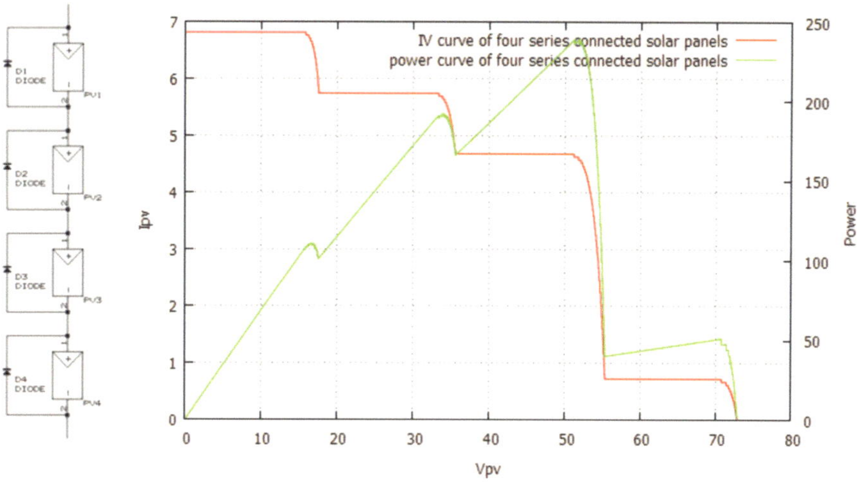

Figure 3. An example of series-connected solar panel property. The irradiations for the four panels are 10, 80, 65, and 95%.

Another noticeable property is that the current flow of photovoltaics stops when it has a shadow. Thus, if a photovoltaic is connected to the other photovoltaics in series and it has a shadow, no power is outputted from the series-connected solar panels.

This problem is solved by connecting a bypass diode in parallel with each photovoltaic. Using this architecture, we can get some amount of power even if a part of the solar panels are under a shadow. However, in such a case, the voltage-power curve of the photovoltaics shows a nonlinear form. As the voltage-power curve has several peaks, the power conditioner cannot obtain the correct MPP only using a hill-climbing technique. The most reliable method to solve this problem is for the power conditioner/DC converter to acquire the current voltage-power curve and detect the global maximum point.

3. MPPT algorithm accelerated by learning machines

One way to realize a quick MPPT without involving any special device is to use a photovoltaic model to predict the MPP. Moreover, the apparent property of photovoltaic varies due to the accumulated dust on the solar panel surfaces. This means that the photovoltaic model is not stable, but is valid depending on the solar panel's situation. To adjust to such changes in the property, an on-site learning machine should learn the MPP acquired by the P&O method to construct the PV model and apply prediction using the learning machine. In our previous work [7], we demonstrated that an incremental learning method on a budget on a microcomputer can manage the learning and prediction of MPPs. The learned results were applied only when solar irradiation changes drastically and the learning machine know the appropriate MPP that fits the current situation.

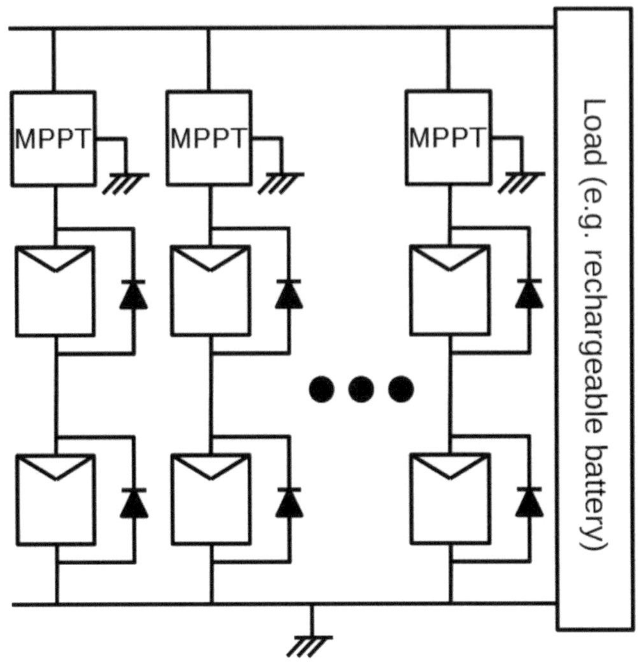

Figure 4. The photovoltaic circuit design bedded on road. Several solar panel strings with the MPPT converter are connected in parallel.

The previous system, however, cannot support the MPPT for series-connected PVs with bypass diodes, as shown in **Figure 4**. This is because even if the strength of solar irradiation is a certain stable value, there are several different solutions depending on the variety of the shadow patterns on the solar panels. To overcome this difficulty, we propose a new MPPT method in this chapter that is based on modal regression on a budget, which is a modal regression with a fixed number of kernels. Modal regression has the ability to approximate multivalued functions. Modal regression on a budget continues the learning with a fixed number of kernels so that it is suitable to be embedded in a small microcomputer. Therefore, it is able to record several different MPPs corresponding to the strength of solar irradiation. The proposed MPPT has a modified P&O method that enables tracking of MPPs from the voltage-power curve having several peaks using modal regression on a budget.

During the service, the proposed MPPT tracks the peaks by changing the initial search points. If an MPP is observed, the kernel density estimator (KDE) in the modal regression records the peak by adding a new kernel that records the current peak (see **Figure 5**). However, the microcomputer has limited storage space. Thus, if the number of kernels in the KDE equals the budget, one of the existing kernels will be replaced by the new kernel.

3.1. A perturbation and observation (P&O) method with changing initial point

Even if the system uses modal regression, it cannot be used before learning. Thus, it needs to obtain the MPPs first. To find several peaks, a modified P&O method is presented. The modified

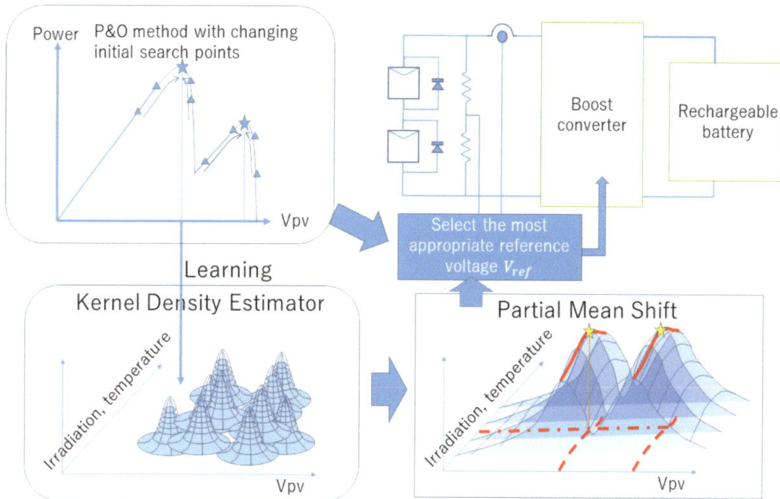

Figure 5. Outline of the MPPT accelerated by the modal regression on a budget.

one searches the peak points roughly at first. For example, if the solar panel comprises m number of clusters, the number of peaks would be up to m. Therefore, the new P&O estimates following C $(> m)$ initial points. This operation concludes when irradiation is greatly changed.

$$v_n^i = n\,\frac{V_{pv}^{max}}{C}, \textit{where } n = 1, \cdots, C, \tag{2}$$

where V_{pv}^{max} denotes the open circuit voltage of the photovoltaic. To obtain this value, the circuit should be opened for a while when the irradiation changes. The system finds the v_{n*}^i that leads to maximum MPP.

$$n^* = argmax_n\{P_{pv}(v_n^i)\}, \tag{3}$$

where $P_{pv}(v_n^i)$ denotes the power from the solar panel for the voltage v_n^i. In this method, m number of clusters are needed to be preset.

3.2. Modal regression on a budget for reasoning from too less sensory inputs

In general, if the device has too few sensors, the system cannot properly detect the current status. The partial shadow problem is one such problem. Therefore, if the device has illuminance sensors for each solar cell, it can accurately detect the status and can form complete relationships between the large number of sensory inputs and MPP. However, such strategy is impractical for real applications. Moreover, we should reduce the number of dimensions to construct an insect's brain like compact learning machine. From a theoretical viewpoint, the system having too few sensory inputs should yield several possible solutions. Therefore, the system has to check the suitability of all possible solutions and choose the best solution. One way to solve this problem is to employ a quick search algorithm such as the PSO algorithm. However, PSO searches possible solutions for arbitrary initial setting of particles and wasted some time for the search. An alternative way to speed up the procedure is by implementing a learning machine to quickly obtain some good solution candidates. However, to realize such tasks, the learning machine has to have an ability to approximate multivalued functions. Such ability cannot be served by normal regression methods.

Modal regression approximates a multivalued function to search the local peaks of a given sample distribution. Modal regression comprises the KDE with a partial mean shift (PMS) method. We have already presented a minimum modal regression, which minimizes the number of kernels for the modal regression [12].

The model, however, does not support learning on a fixed budget. In this chapter, we propose an improved version of our previous work, which enables learning on a fixed budget.

3.2.1. *Original modal regression method*

Modal regression comprises KDE followed by the PMS. KDE is a variation of the Parzen window [13]. Let $\aleph$ be the set of learning samples and $\aleph = \{x_p \in \Re^N | p = 1, 2, \ldots N\}$. The estimator approximates the probability density function using a number of kernels, namely the support set S_t. The kernels used are Gaussian kernels and

$$p(\boldsymbol{x}) \propto \sum_{i \in S_t} K\left(\frac{\|\boldsymbol{x} - \boldsymbol{x}_i\|}{h_x}\right) \tag{4}$$

where

$$K\left(\frac{\|\boldsymbol{x} - \boldsymbol{x}_i\|}{h_x}\right) \equiv exp\left(-\frac{\|\boldsymbol{x} - \boldsymbol{x}_i\|^2}{h_x^2}\right) \tag{5}$$

Normally, the same number of kernels as that of the dataset is required. However, if the storage capacity of a target device is small, the number of kernels must be restricted. There are several ways to realize density estimation using a limited number of kernels. Traditionally, self-organizing feature maps or learning vector quantization methods approximate the distribution using a fixed number of templates.

As mentioned in a previous study [14], the KDE used in modal regression should approximate the number of peak points of the distribution, rather than the distribution itself. Let $\widehat{p}(\mathbf{x})$ be

$$\widehat{p}(\boldsymbol{x}) \equiv \sum_{i \in S_t} K\left(\frac{\|\boldsymbol{x} - \boldsymbol{x}_i\|}{h_x}\right) \tag{6}$$

then $\widehat{p}(\mathbf{x})$ should satisfy the following condition.

$$\begin{cases} \nabla_x \hat{p}(\boldsymbol{x})\Big|_{\boldsymbol{x}=\boldsymbol{x}^*} = \nabla_x p(\boldsymbol{x})\Big|_{\boldsymbol{x}=\boldsymbol{x}^*} = 0 \\ \nabla_x^2 \hat{p}(\boldsymbol{x})\Big|_{\boldsymbol{x}=\boldsymbol{x}^*} < 0, \ \nabla_x^2 p(\boldsymbol{x})\Big|_{\boldsymbol{x}=\boldsymbol{x}^*} < 0 \end{cases}. \tag{7}$$

where $\mathbf{x}^*$ denotes a local peak point of the distribution.

Modal regression searches the peaks of the distribution model represented by the KDE. The PMS method realizes quick convergence to the nearest peak from the initial point. Let us denote the initial point as $\mathbf{x}_0$, representing the starting point for searching the peaks. Thus, modal regression repeats the modification of the current y as follows:

$$y_{new} \leftarrow \frac{\sum_i y_{old} K\left(\frac{|y_{old}-y_i|}{h_y}\right) K\left(\frac{\|\boldsymbol{x}-\boldsymbol{x}_i\|}{h_x}\right)}{\sum_j K\left(\frac{|y_{old}-y_j|}{h_y}\right) K\left(\frac{\|\boldsymbol{x}-\boldsymbol{x}_j\|}{h_x}\right)} \tag{8}$$

3.2.2. *Modal regression on a fixed budget*

To embed the modal regression, we have to pay attention to how to reduce the number of kernels for the KDE. Especially, we have to fix the upper bound for the number of kernels. In this case, the aim of the KDE is to approximate the peaks in the distribution rather than approximating the distribution. From this viewpoint, we should prune redundant kernels that do not contribute to approximating the peaks.

In our previous work [12], we demonstrated that the kernel, which is linearly dependent on the other kernels, can be removed without changing existing peaks. To this end, before pruning, the pruned kernel should be projected to the space spanned by the other remaining kernels. However, preparing the gram matrix wastes huge memory space.

Moreover, in this practical application, we should pay attention to the concept drift phenomena, wherein the labels change over time. This is caused by environmental changes such as the accumulation of dust on the solar panels and the changes in properties of the solar panel materials. The learning methods should support these issues.

To overcome these difficulties, we propose a simplified version of the modal regression method on a fixed number of kernels.

To discuss the learning rule of the KDE, let us rewrite the kernel output value as the dot product of the two vectors of $\mathrm{k}(\boldsymbol{x}_i, .)$ and $\mathrm{k}(\boldsymbol{x}, .)$ as follows.

$$\langle k(\boldsymbol{x}_i, \cdot), k(\boldsymbol{x}, \cdot)\rangle \equiv K\left(\frac{\|\boldsymbol{x}-\boldsymbol{x}_i\|}{h_x}\right), \tag{9}$$

where $\langle \cdot, \cdot \rangle$ denotes the dot product operator. This expression is based on the kernel method. Fortunately, the Gaussian kernel is a type of reproducing kernel in which we can rewrite the learning rule using the dot product of vectors. Using this representation, we can rewrite the learning rule in algebraic expressions, which can be very easily understood. Now, let us denote a vector $\widehat{P}_t$ as the learning result after the t-th sample presentation. Then, we obtain

$$\widehat{P}_{t-1} \equiv \sum\nolimits_{i \in S_{t-1}} W_i k(\boldsymbol{x}_i, \cdot), \tag{10}$$

where S_{t-1} denotes the support set after the $t-1$-th presentation of a given sample. The KDE output to an input vector $\boldsymbol{x}$ is calculated by

$$\widehat{P}_{t-1}(\boldsymbol{x}) = \left\langle \widehat{P}_{t-1}, k(\boldsymbol{x}, \cdot) \right\rangle. \tag{11}$$

Eq. (10) enables us to represent the learning rule as

$$\widehat{P}_t = \widehat{P}_{t-1} + y_t k(\boldsymbol{x}_t, \cdot), \; S_t = S_{t-1} \cup \{t\} \tag{12}$$

However, the proposed method restricts the number of kernels to a certain number as $|S_t| \leq B$. To overcome this problem, the proposed method replaces one of the kernels with a new kernel whose centroid is the new input vector, or moves the nearest kernel centroid to close to the current new input vector. Therefore, if the nearest kernel

$$n_t = argmin_j \left\{ \|\boldsymbol{x}_t - \boldsymbol{x}_j\|^2 \right\}, \tag{13}$$

satisfies the following condition

$$\|\boldsymbol{x}_t - \boldsymbol{x}_{n_t}\|^2 < \theta_{activity}, \tag{14}$$

its kernel center is modified to be the mean vector of the original kernel center and the new sample as follows. The extension coefficient W_{n_t} is increased by Δ.

$$\mathbf{x}_{\mathrm{n_t}} = \frac{\left(\frac{\mathrm{W}_{\mathrm{n_t}}}{\Delta}\right)\mathbf{x}_{\mathrm{n_t}} + \mathbf{x}_{\mathrm{t}}}{\left(\frac{\mathrm{W}_{\mathrm{n_t}}}{\Delta}\right) + 1}, \quad \mathrm{W}_{\mathrm{n_t}} = \mathrm{W}_{\mathrm{n_t}} + \Delta \tag{15}$$

The extension coefficient includes information on how many samples did the kernel learn. The extension coefficient is also reflected to a weighted PMS method in Eq. (20). However, if the kernel center does not satisfy the Eq. (14), one of the kernels should be replaced with the new tentative kernel. Therefore, if the new sample $\mathbf{x}_t$ is too far from the nearest kernel center, one of the kernels should be replaced with it to adjust to the new sample. In such a case, the least recently or frequently used (LRFU) kernel is to be replaced with the new one. The LRFU evaluation method proposed in [15] is an improved version of the LRU page-replacement algorithm for virtual memory systems on operating systems. Using this evaluation method, the most ineffective kernel, which seems to be unused for a long time interval, is replaced with the new kernel. To realize this evaluation, a variable that represents the value of each kernel is introduced. Let C_i be the value of the i-th kernel. When the i-th kernel centroid is the closest to current sample x_t, C_i is enlarged, but is decreased, otherwise. Therefore, for each round, the following equation is executed.

$$C_i = \begin{cases} C_i + 1, & i = n_t \\ \eta C_i, & i \neq n_t \end{cases}, \tag{16}$$

where $\eta = 1 - \epsilon,\ \ \epsilon \ll 1$. Then, the $j^* = argmin_i C_i$ kernel is to be replaced with the new kernel. Therefore,

$$\boldsymbol{x}_{j^*} = \boldsymbol{x}_t,\ \ w_{j^*} = \Delta,\ \ C_{j^*} = 1,\ \ y_{j^*} = y_t \tag{17}$$

h_x determines the width of each kernel. The performance of the system is also sensitive to this value, so we have to set this value carefully. In a previous study [16], the optimal value of h_x for a standard distribution was derived as

$$\mathrm{h_x} = \left(\frac{4}{d+2}\right)^{\frac{1}{d+4}} n^{-\frac{1}{d+4}}, \tag{18}$$

where $\mathrm{d} = \dim(\boldsymbol{x}_t)$ is the dimension of the input vector and n is the number of samples. In this study, the number of samples is unknown. However, the number of kernels are bounded to the budget B so that $\mathrm{n} = \mathrm{B}$. Equation (18), however, cannot be used for practical applications. Therefore, we should consider a scaling factor for (18). To this end, in this study, we rewrite (18) as follows.

$$\mathrm{h_x} = v_0 \cdot \left(\frac{4}{d+2}\right)^{\frac{1}{d+4}} n^{-\frac{1}{d+4}}, \tag{19}$$

where v_0 denotes the scaling factor and was set to 0.3 in this simulation described in Section 4. Actually, in the simulation described in Section 4, each input dimension was normalized before the execution of the modal regression. Concretely, each element of $\boldsymbol{x}_t$ of modal regressor was multiplied by a gain g_i to make the range of the ith element of $\boldsymbol{x}_t$ be $|g_i x_{ti}| \leq 1$. The output from the modal regressor (20) was divided by the corresponding gain $y = y/g_o$. For simplicity, however, following text omit the description of these gains.

The regression output is also delivered by the PMS method described in (8). In this model, the PMS method should account for the extension parameter $\mathrm{W_i}$. To this end, this method also uses the weighted PMS method as is done in our previous work [12]. Note that (20) includes the extension parameter $\mathrm{W_i}$ in both the numerator and the denominator.

$$y_{new} \leftarrow \frac{\sum_i y_{old} W_i K\left(\frac{|y_{old} - y_i|}{h_y}\right) K\left(\frac{\|\boldsymbol{x} - \boldsymbol{x}_i\|}{h_x}\right)}{\sum_j W_j K\left(\frac{|y_{old} - y_j|}{h_y}\right) K\left(\frac{\|\boldsymbol{x} - \boldsymbol{x}_j\|}{h_x}\right)} \tag{20}$$

The weighted PMS should be repeated by substituting derived $\mathrm{y_{new}}$ to $\mathrm{y_{old}}$ until it converges to a certain value. In the computer simulation described in Section 4, the weighted PMS was repeated 10 times for every initial point. This process is executed for all initial values of y_{old} to obtain all local peaks. The simplest way to set the initial points is choosing uniform random initial values for $\mathrm{y_{old}}$. However, the random initial values usually make some unexpected

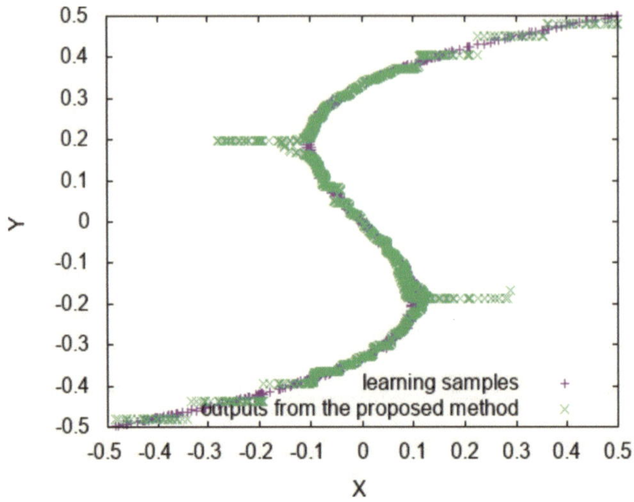

Figure 6. The response for the third-order data. The green curve is the response of the proposed model with 50 kernels.

converged values for y. To more appropriately set up the initial value y_0, the proposed method chooses the initial value as the corresponding element of each kernel center. Therefore, let us assume that a kernel center $\mathbf{x}_i$ is similar to the current input. Then, the initial value should be $y_0 = x_{ij}$, where j is the corresponding unknown dimension. The set of such kernel centers is

$$S_{active} \equiv \left\{ i \middle| exp\left(\sum\nolimits_{j \neq unknown} \frac{-\left(x_{ij} - x_{tj}\right)^2}{h_x^2} \right) > \theta_{init} \right\}, \tag{21}$$

where θ_{init} denotes the threshold for choosing the kernel. The above equation does not contain the distance calculation for the unknown dimension. The initial values for $y_0$are

$$y_0 = x_{k\ unknown} \ \ where\ k \in S_{init}. \tag{22}$$

3.2.3. *An example of the modal regression outputs*

The modal regression approximates multivalued functions. As an example, **Figure 6** shows the regression output for 800 sets of third-order synthetic data with 50 kernels. We can observe that the proposed method partly approximates multivalued function.

3.3. Whole algorithm

Algorithms 1–4 are presented below. Note that S_t in these algorithms shows the averaged solar irradiation for all clusters. Therefore, solar irradiation is assumed to be sensed by a single illuminance sensor; thus, the obtained value is the average of the values of both clusters.

The algorithm is roughly divided into two parts: one is the normal P&O part, and the other deals with searching for the reference voltage using the proposed modal regression. The second part is executed when the solar irradiation is changed abruptly.

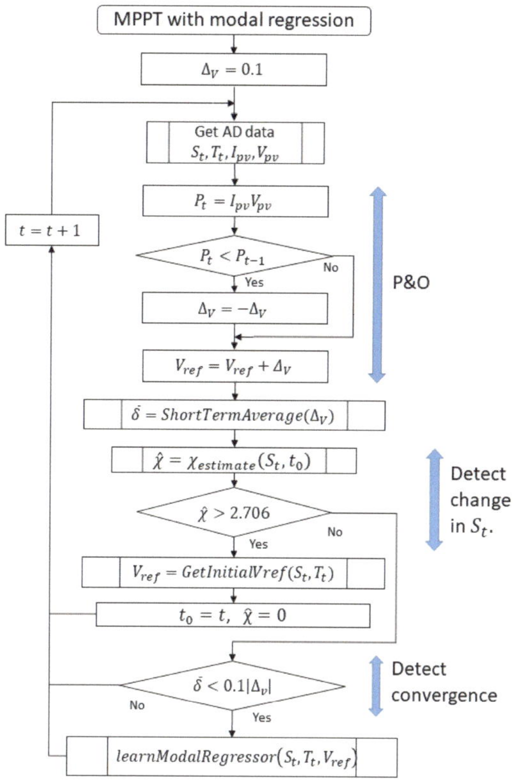

Algorithm 1. Algorithm of the MPPT with modal regression. Note that V_{ref} is referenced by the proportional-integral-derivative (PID) control thread at each time interval. *GetInitialVref()* is described in **Algorithm 2**. *learnModalRegressor*() is described in Eqs. (12)–(19).

To this end, the sensed solar irradiation is statistically analyzed by $\chi_{\mathbf{estimate}}(S_t, t_0)$, that is chi-square test for $\mathbf{S_t}$ from the previously changed time $\mathbf{t_0}$ till now. If it includes an obvious change, **Algorithm 2** is called to search for an appropriate initial V_{ref}'s for searching the optimal value of V_{ref}. **Algorithm 2** is the algorithm for searching initial V_{ref}. This algorithm conducts a search using the proposed modal regressor followed by a search of the initial V_{ref} using the proposed modified P&O algorithm described by Eq. (2). The reason why it executes an additional search is that there is a possibility that the modal regressor yields incomplete solution candidates. Such a situation usually occurs when the modal regressor is in the early stage of learning.

3.4. Computational cost and required memory capacity

The computational cost for the MPPT with modal regression is mainly wasted by the modal regressor. Hence, let us consider the computational cost for the modal regression. Now, we assume that the number of kernels in the modal regressor is $\boldsymbol{B}$ and that the number of dimensions is $\boldsymbol{N}$. Note that $\boldsymbol{N}$=**3** because input vector is $\boldsymbol{x}_t = \left[\boldsymbol{S}_t, \boldsymbol{T}_t, \boldsymbol{V}_{ref}\right]^T$. To calculate the kernel outputs for current input $\boldsymbol{x}_t$, it needs **(N+1)**B times multiplies and $\boldsymbol{B}$ times of division and $\boldsymbol{B}$ times of calculation of **exp** (). If we assume that the calculation of **exp** () is $\boldsymbol{C}_{exp}$, the

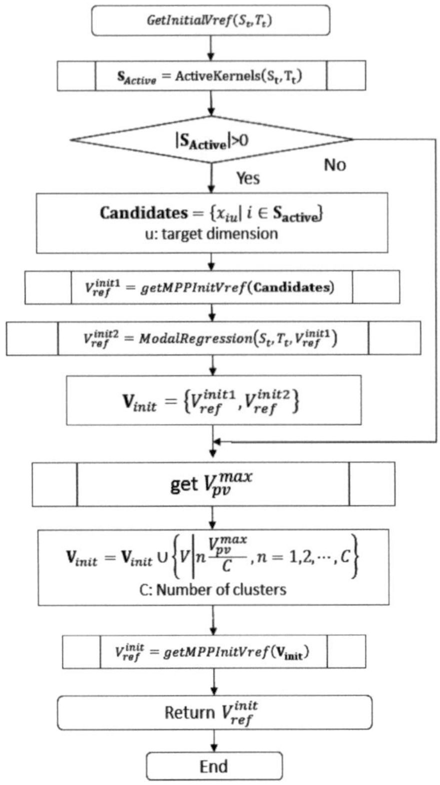

Algorithm 2. Pseudo code for getting initial reference voltage. *ActiveKernels*() is derived by (21). *getMPPInitVref*() is described in **Algorithm 3**. *ModalRegression*() is the five time repeats of the partial mean shift:(8).

computational cost is proportional to $B(N+2+C_{exp})$. Therefore, to derivate a kernel set S_{active} in (21), it needs $O(B)$. The partial mean shift (20) needs $B(N+8+2C_{exp})+B(N+7+2C_{exp})+1=B(2N+15+4C_{exp})+1$. Thus, if the partial mean shift is repeated for M times for each trial, the total computational power of modal regression is proportional to $MB(2N+15+4C_{exp})+1$.

The computational power required for the learning of the modal-regressor is the cost of executing (13), (14), and (16). Thus, it needs $BN+N+(2N+1)$ multiplications. After all, the computational complexity of the modal regression is $O(B)$.

The required memory capacity also depends on the number of kernels. Each kernel records the center of kernel x_i, corresponding label y_i, the extension parameter W_i and the parameter C_i for the LRFU estimation. As each float variable requires 4 bytes, one kernel requires $4(N+2)$ bytes. Thus, the total amount of memory storage for all kernels is $4B(N+2)$ bytes.

The boost converter step ups the voltage of the solar panel string and charges the battery. The MPPT unit, which includes the proposed method, sends the predicted MPP: V_{ref} to the feed-back controller. The P-type MOSFET is assumed to be used for making an open circuit in a short-time interval to get V_{pv}^{max} (see **Algorithm 2**). As shown in **Figure 4**, several sets of this circuit are connected in parallel to the same rechargeable battery.

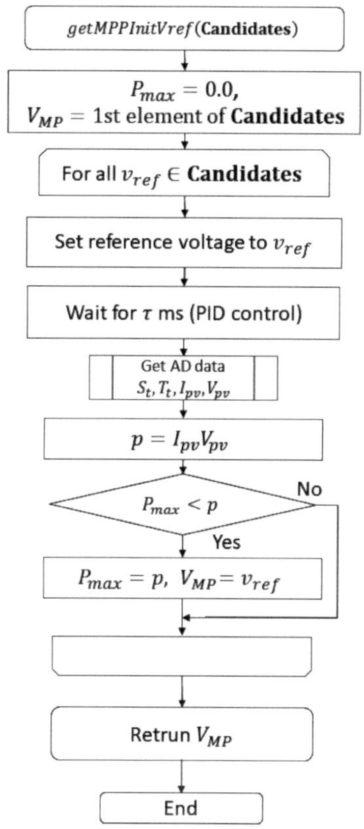

Algorithm 3. Flowchart for *getMPPInitVref*().

4. Computer simulation

The performance of the proposed MPPT was evaluated via a simulation. Particularly, the convergence speed to MPP is a very important property that should be evaluated. The simulated circuit comprises a short string of solar panels connected to a boost converter (see **Figure 7**).

The MPPT unit sends the reference voltage V_{ref} for the feedback controller, and the boost chopper circuit adjusts the output voltage of the PV string to V_{ref}. In this simulation, we assume that the load is a rechargeable battery, whose voltage is kept to a certain constant voltage. Using this load, each boost converter is not affected by the change in the other converter's output power.

For simplicity, the simulator of the boost converter simply updates V_{pv} to be V_{ref} and calculates the corresponding I_{pv} by using the photovoltaic model. Therefore, the detailed transient response of the boost converter was not realized in the simulator.

To realize the simulation, we constructed a simulator of photovoltaics and circuits as the Java application. The solar irradiation, temperature, and the properties of the solar panels are also represented in the thread of environment class (see **Figure 8**).

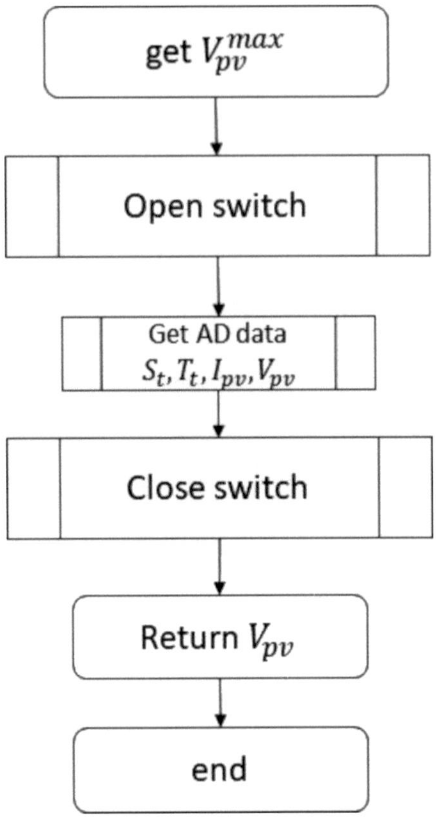

Algorithm 4. Flowchart for getting V_{pv}^{max}. Open and close switch denote enabling and disabling the FET in **Figure 7**.

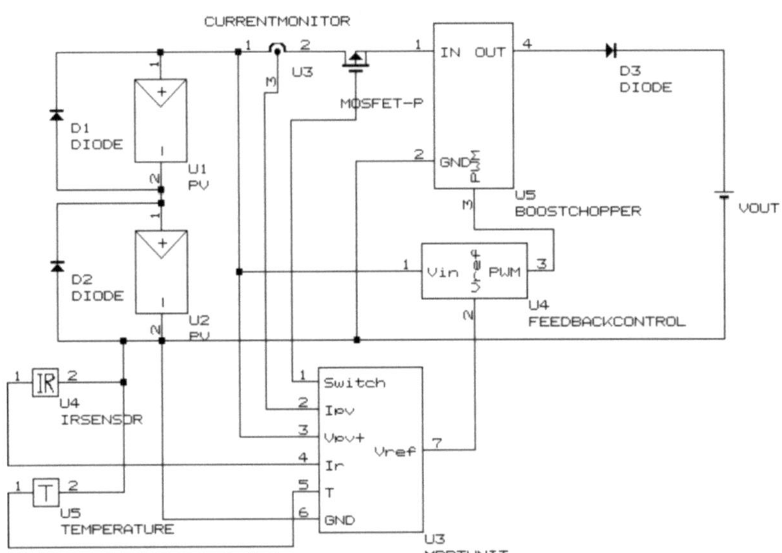

Figure 7. The circuit for the simulation.

http://dx.doi.org/10.5772/intechopen.79711

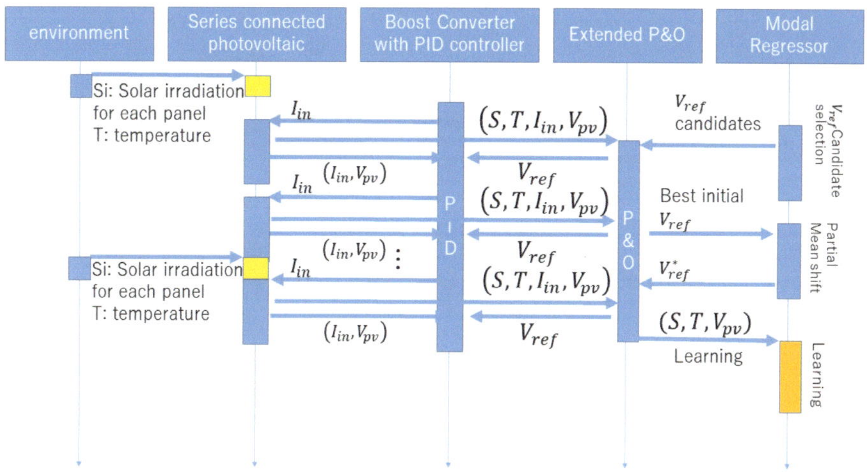

Figure 8. Sequence diagram of the simulator.

For simplicity, the strength of solar irradiation and temperatures varies for a certain scenario, but the effect of the specific heat of the solar panel material was not considered.

The solar panel is a homogeneous two cluster panel such that it has two peaks under partial shadow conditions. The MPPT with modal regression is also represented by the MPPT thread class. The chopper circuit with the feedback controller is assumed to control the output voltage from the solar panel to V_{ref}, which is assigned by the MPPT unit, within 1 ms. Note that V_{ref} is yielded by the modified P&O method or the modal repressor. Similar to the simulation method proposed in [10], the chopper circuit is simulated so as to change I_{pv}. As a result, the series-connected solar panel simulator yields a new V_{pv} due to the change in I_{pv}. The new V_{pv} is then sent to the boost converter simulator to calculate the next step.

We have compared the proposed method with the existing models under partial shadow conditions. For this comparison, the following three models were prepared: MPPT with the modal regression, the P&O method by changing initial points described in Section 3.1, and MPPT with PSO. There are various PSO-based MPPT methods [8, 17]. In this simulation, we prepared a model that is based on the model proposed in [17] because it has a similar

Δ_V: Change in voltage for P&O (**Algorithm 1**)	0.1
θ_{init} in Eq. (21)	0.9
η in Eq. (16)	0.001
d in Eq. (19)	3
Time interval for changing V_{ref} by P&O, modal regression and PSO (=τ in **Algorithm 3**)	1 [ms]
Time interval for changing solar irradiation	250 [ms]
Scaling factor v_0 in (19)	0.3
Number clusters (= C in **Algorithm 2**). This value should be greater than the actual number of clusters.	3

Table 1. Parameters used in this simulation.

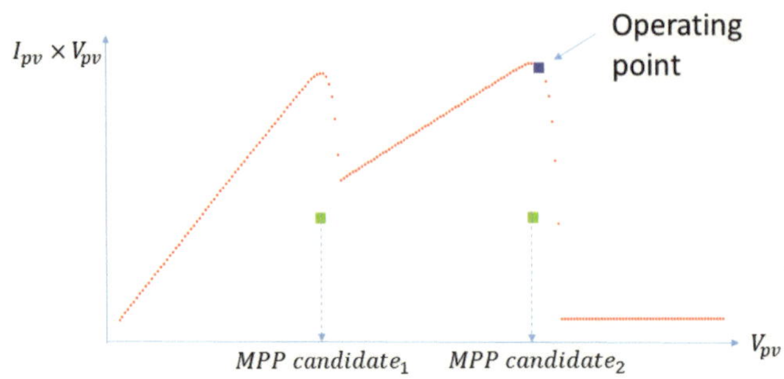

Figure 9. An example of snapshot of the maximum power tracking of the proposed method. The green points are the center points of the proposed modal regressor, namely the initial MPP candidates (see (21)).

architecture to ours. The PSO-based MPPT method used in this simulation executes the PSO optimization when solar irradiation changes is occurred. The condition for detecting solar irradiation changes was the same as the method described in Section 3.3. The detailed parameters used in this simulation are listed in **Table 1**.

We evaluated the electric power generation behavior of each model. If the generated power is higher than the others, the model finds MPP faster than the others.

Figure 9 shows a snapshot of the behavior of our proposed MPPT. In this situation, the power-voltage curve of the solar panel has two peak points. The activated kernel centers of the modal-regression at this situation are shown as the two green points[1]. The proposed method set choose one of them as the start point for the MPPT. After that, the modal regression output was used for the

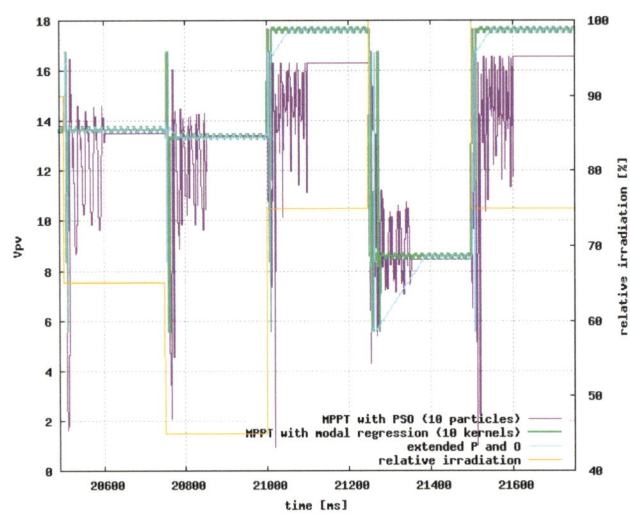

Figure 10. An example of V_{pv} VS time.

[1]The activated kernel centroids without the power element were pointed as the green points. However, the height of the green points have been set to a certain fixed value for easy seeing.

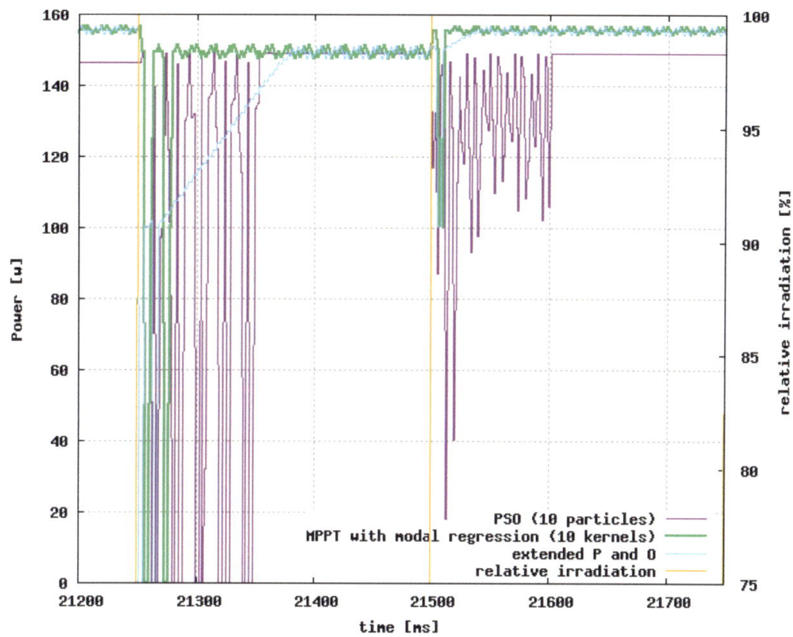

Figure 11. The magnified power curves. Note that the power curve has changed immediately after the change of irradiation.

initial point for starting the P&O procedure. As a result, the proposed method finds the MPP faster than the P&O method. The quick search ability is suitable for generating electricity under changing irradiation. In **Figure 10**, the green, blue, and purple curves show the V_{pv} of the proposed one, P&O-, and PSO-based MPPT methods, respectively. The V_{pv} of the PSO-based method changes drastically for approximately 100 ms immediately after the change in solar irradiation. Although the proposed and P&O methods also change V_{pv} immediately after the change in solar irradiation, the changing period is shorter than that of the PSO-based method. Moreover, V_{pv} of P&O-based method sometimes needs a time interval to converge to be a steady state. On the contrary, the proposed method makes V_{pv} reach the steady state faster than the others.

The magnified **Figure 11** shows that the power generation of our proposed method quickly aliased immediately after the change in solar irradiation, whereas the extended P&O method gradually converges to the power of the proposed method. The PSO based MPPT shows the less power generation than the other methods. In the case of PSO, the results are greatly affected by the initial points of the particles. In this simulation, we have set the initial points by uniform random voltages in $[0, V_{PV}^{max}]$, where V_{PV}^{MAX} is the open-circuit voltage of the solar panel string. The initial points should be distributed uniformly in the interval. However, if the number of particles is small due to the restriction of the device, the initial point distribution usually becomes to be an unbalanced distribution. As a result, the quality of the solution is degraded. To check the performances under the various sizes of kernels or particles, the averaged generated power for the proposed method with 5 and 10 kernels, and the PSO-based MPPT methods with 5 and 10 particles were compared. Moreover, the generated electricity power from the proposed method and the extended P&O

Time interval for changing solar insolation	Method	Averaged electricity power
250 ms	MPPT with modal regression (5 kernels)	151.2 W
	MPPT with modal regression (10 kernels)	151.9 W
	Extended P&O	150.3 W
	MPPT with PSO (5 particles)	127.1 W
	MPPT with PSO (10 particles)	128.2 W
200 ms	MPPT with modal regression (10 kernels)	151.4 W
	Extended P&O	149.4 W

Table 2. Comparison of averaged electricity power generated during the first 200 [s]. The time interval for solar irradiation change were 250 and 200 [ms].

methods were compared with two different time intervals of changing solar irradiation. **Table 2** shows the results. We can see that the averaged generated power of the proposed method of 5 and 10 kernels are almost the same. On the other hand, the PSO-based methods reduced the power if the size of particles is reduced. The proposed method's generated power was also larger than the extended P&O method because the convergence speed is higher than that of the P&O method. The difference in the generated power is caused by their different convergence speed. Therefore, if there are fewer changes in solar irradiation, the difference decreases because the convergence process does not occur. As evidence, **Table 2** shows that if the time interval of changing solar irradiation is 250 ms, the differences between the two averaged generated power was 1.6 W, whereas the difference was 2 W when the time interval is 200 ms.

5. Conclusion

In this chapter, we proposed a new MPPT method accelerated by modal regression on a budget, which approximates multivalued functions. The modal regression on a budget is a simplified version of our previously proposed method, namely limited modal regression [12].

The proposed MPPT method comprises an irradiation sensor, temperature sensor, and modal regression on a budget. We assume that the irradiation sensor gets the averaged strength of irradiation of all solar panels. In the case for MPPT of PV strings, the device has to obtain the highest local peak point from the several peak points in the voltage-power curve. Therefore, the MPPT device with the incomplete sensory input has to approximate a multivalued function between the sensory inputs and the MPP.

Normally, modal regression estimates provide sample distribution and yield local peak points that are related to the specified input.

The modal regression on a budget can approximate such relationships between the sensory inputs and the MPP's. The proposed MPPT method is a combination of modal regression on a budget and a modified (extended) P&O method. The modified P&O method obtains the MPPs even if there are several local peak points. The obtained MPPs are recorded in the modal regressor.

The proposed method was evaluated by computer simulation under partial shadow conditions. The simulation results suggest that the MPPT with modal regressor obtain an MPP faster than other existing methods such as the MPPT with PSO. This property is suitable for electricity generation using the solar panels bedded on roads.

Acknowledgements

This study is sponsored and supported by KYODO Corporation, Toyota-city, Aichi-ken.

Author details

Koichiro Yamauchi

Address all correspondence to: k_yamauchi@isc.chubu.ac.jp

Department of Computer Science, Chubu University, Kasugai-shi, Aichi, Japan

References

[1] Bendib B, Belmili H, Krim F. A survey of the most used mppt methods: Conventional and advanced algorithms applied for photovoltaic systems. Renewable and Sustainable Energy Reviews. 2015;**45**:637-648

[2] Esram T, Chapman PL. Comparison of photovoltaic array maximum power point tracking techniques. IEEE Transactions on Energy Conversion. 2007;**22**(2):439-449

[3] Boehringer AF. Self-adapting dc converter for solar spacecraft power supply. IEEE Transactions on Aerospace and Electronic Systems. 1968;**AES-4**(1):102-111

[4] Esram T, Kimball JW, Krein PT, Chapman PL, Midya P. Dynamic maximum power point tracking of photovoltaic arrays using ripple correlation control. IEEE Transactions on Power Electronics. 2006;**21**(5):1282-1291

[5] Veerachary M, Senjyu T, Uezato K. Neural-network-based maximum-power-point tracking of coupled-inductor interleaved-boost-converter-supplied pv system using fuzzy controller. IEEE Transactions on Industrial Electronics. 2003;**50**(4):749-758

[6] Akkaya R, Kulaksiz AA, Aydogdu O. Dsp implementation of a pv system with ga-mlp-nn based mppt controller supplying bldc motor drive. Energy Conversion and Management. 2007;**48**:210-218

[7] Yamauchi K. Incremental learning on a budget and its application to quick maximum power point tracking of photovoltaic systems. Journal of Advanced Computational Intelligence and Intelligent Informatics. 2014;**18**(4):682-696

[8] Liu C-L, Luo Y-F, Huang J-W, Liu Y-H. A pso-based mppt algorithm for photovoltaic systems subject to inhomogeneous insolation. In: The 6th International Conference on Soft Computing and Intelligent Systems. 2012. pp. 721-726

[9] Noguchi T, Togashi S, Nakamoto R. Short-current pulse-based maximum-power-point tracking method for multiple photovoltaic-and-converter module system. IEEE Transactions on Industrial Electronics. 2002;**49**(1):217-223

[10] Tan YT, Kirschen DS, Jenkins N. A model of pv generation suitable for stability analysis. IEEE Transactions on Energy Conversion. 2004;**19**(4):748-755

[11] Bellia H, Youcef R, Fatima M. A detailed modeling of photovoltaic module using matlab. NRIAG Journal of Astronomy and Geophysics. 2014;**3**:53-61

[12] Koichiro Y, Vanamala Narashimha B. Minimum modal regression. In Maria De Marsico, Gabriella Sanniti di Baja, and Ana Fred, editors, ICPRAM2018 7th International Conference on Pattern Recognition Applications and Methods, pages 448-455, 2018

[13] Parzen E. On estimation of a probability density function and mode. Annals of Mathematical Statistics. 1962;**33**(3):1065-1076

[14] Sasaki H, Ono Y, Sugiyama M. Modal regression via direct log-density derivative estimation. In: Hirose A, Ozawa S, Doya K, Ikeda K, Lee M, Liu D, editors. Neural Information Processing –23rd International Conference, ICONIP 2016–, Volume Part II. Springer-Verlag; 2016

[15] Lee D, Noh SH, Min SL, Choi J, Kim JH, Cho Y, Sang KC. Lrfu: A spectrum of policies that subsumes the least recently used and least frequently used policies. IEEE Transactions on Computers. 2001;**50**(12):1352-1361

[16] Silverman BW. Density Estimation for Statistics and Data Analysis. CRC Press; 1986

[17] Rajasekar N, Vysakh M, Thakur HV, Mohammed Azharuddin S, Muralidhar K, Paul D, Jacob B, Balasubramanian K, Sudhakar Babu T. Application of modified particle swarm optimization for maximum power point tracking under partial shading condition. Energy Procedia. 2014;**61**:2633-2639

Chapter 6

Optimal Designing Grid-Connected PV Systems

Ali Reaz Reisi and Ashkan Alidousti

Additional information is available at the end of the chapter

http://dx.doi.org/10.5772/intechopen.79685

Abstract

Photovoltaic systems, direct conversion of solar energy to electrical energy, are produced in the form of DC power by photovoltaic arrays bathed in sunlight and converted into AC power through an inverter system, which is more convenient to use. There are two main paradigms for optimal designing of photovoltaic systems. First, the system can be designed such that the generated power and the loads, that is, the consumed power, match. A second way to design a photovoltaic system is to base the design on economics, as pinpointed in the following. Photovoltaic grid connected through shunt active filter by considering maximum power point tracking for these systems is known as the optimal design. This chapter is organized as follows: First, we discuss an overview of grid-connected photovoltaic systems. After that, we take a more detailed look on grid-connected photovoltaic system via active filter; in this section, we explain the modeling of photovoltaic panel and shunt active filter. In the next section, we learn different maximum power point tracking methods and also learn how to design DC link as a common bus of shunt active filter and photovoltaic system. Finally, MATLAB/Simulink simulations verify the performance of the proposed model performance.

Keywords: optimal designing, grid connected, photovoltaic systems, shunt active filter, maximum power point tracking

1. Introduction

Global warming, environmental pollution, and possible scarcity of fossil fuel reserves are some of the main driving forces behind the urge for installing grid-connected photovoltaic (PV) systems. Moreover, utilities and customers can benefit from installing these systems. The main gain for customers is to take advantage of the incentives provided by the governments upon installing PV systems. For utilities, the gains of installing PV systems are mainly operational benefits, especially if the PV system is installed at the customer side on rural feeders. For example, PV systems can be used to decrease the feeder losses, improve the voltage profile of the feeder, and reduce the lifetime operation and maintenance costs of transformer load tap changers. Moreover, if the peak output of the PV system matches the peak loading of the feeder, then the loading of some transformers present in the network can be reduced during peak load periods. The power-quality index of the grid could be improved if PV system was connected to grid via active filters; they are energy conditioners, which include DC/AC-controllable converters. These filters, based on their control schemes, can compensate both source voltage deficiencies and undesirable load-terminal current, which leads to have a source end purely sinusoidal current.

Current harmonic drawn by direct PV grid connected via DC/AV inverters and also nonlinear loads disturb the waveform of the voltage at the point of a common coupling (PCC) and lead to voltage harmonics. Therefore, it is necessary to develop techniques to reduce all the harmonics as it is recommended in the IEEE 519–1992 standard [1]. The first approach consists of the design of LC filters. However, passive filters are not well adapted as they do not take into account the time variation of the loads and the network. They can also lead to resonance phenomena. So, since several years, a more interesting technique is studied: the active filter based either on voltage source or on current source inverters, yielding the harmonic currents required by the load.

Recently, efforts [1, 2] have been made to combine the active filters with renewable energy production systems to benefit from advantages of both a renewable source of energy and a power conditioner to provide pollution-free and high-quality power to the consumers. It seems that it is necessary to use power conditioners, for example, active filters, to compensate the power-quality problems caused by renewable sources of energy in some cases. Since passive filter like LC filters lack the capability to fully compensate these problems in the presence of nonlinear loads, they are not preferred rather than active filters. This type of modular and renewable technology has many advantages like the capability to be expanded and being practically applied in almost everywhere. Furthermore, since one less converter is used in this scheme, there will be capital investment saving in comparison with a separated shunt active filter (SAF) and a PV system; since here by a common DC bus for both SAF and PV systems is used, the cost of the produced power will be reduced.

2. Models of PV and SAF

2.1. PV model

The equivalent circuit of a PV cell is presented in **Figure 1**. Photovoltaic cells of a solar panel have three kinds of external connections, namely series, parallel, and series-parallel. Eq. (1) presents voltage-current characteristic of a solar panel [1], and I_{pv} and I_o are calculated based on the following:

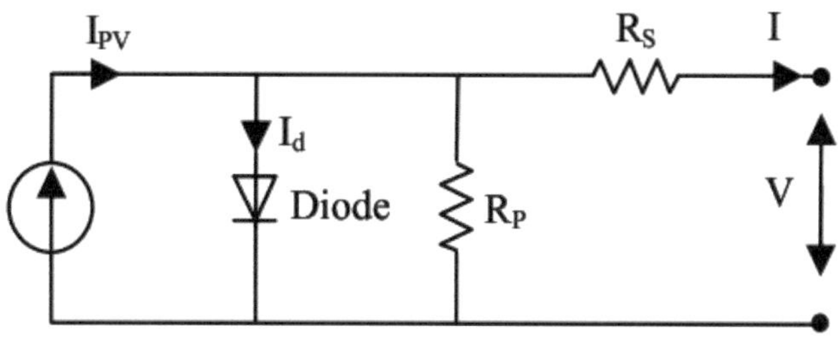

Figure 1. Equivalent circuit of a PV cell.

$$I = I_{PV} - I_O\left[\exp\left(\frac{V + R_S I}{aV_t}\right) - 1\right] - \frac{V + R_S I}{R_p} \tag{1}$$

$$I_{PV} = (I_{PV,n} + K_I \Delta T)\frac{G}{G_n} \tag{2}$$

$$I_O = \frac{I_{SC,n} + K_I \Delta T}{\exp(V_{OC,n} + K_V \Delta T)/aV_t - 1} \tag{3}$$

where I_{PV} and I_o are the photovoltaic current and saturated reverse current, respectively, while "a" and "K" are the ideal diode constant and Boltzmann constant, respectively. Also, we have $V_t = N_S KTq^{-1}$ which is the thermal voltage, N_S is the number of series cells, q is the electron charge, and T is the temperature of p-n junction. R_S and R_P are series and parallel equivalent resistance of the solar panels, respectively. I_{PV} is varied with light intensity in a linear relation and also varies with temperature variations. I_O is dependent on temperature variations.

We have $I_{PV,n}$, $I_{SC,n}$, and $V_{OC,n}$ which are photovoltaic current, short-circuit current (SCC), and open-circuit voltage (OCV) in standard conditions (T_n = 25°C and G_n = 1000 Wm^{-2}), respectively. K_I stands for the coefficient of short-circuit current to temperature, $\Delta T = T - T_n$presents the temperature deviation from standard temperature, G is the light intensity, and K_V shows the ratio coefficient of open-circuit voltage to temperature.

Three important points of I-V characteristic of solar panels are open-circuit voltage, short-circuit current, and voltage-current corresponding to the maximum power. The mentioned points are varied by changes in atmospheric conditions. Short-circuit current and open-circuit voltage can be calculated in different atmospheric conditions by using Eqs. (4) and (5) which are derived from PV model equations, as follows:

$$I_{SC} = (I_{SC,n} + K_I \Delta T)\frac{G}{G_n} \tag{4}$$

$$V_{OC} = V_{OC,n} + K_V \Delta T \tag{5}$$

2.2. SAF

SAF is used to eliminate load-terminal current harmonics and consequently having a pure sinusoidal source-end current. The Generalized Theory of Instantaneous Power (GTIP) theory, as control algorithm, is used for generating reference signal in the activating algorithm of the shunt active filter [3].

U(t) is assumed as the load voltage which consists of all voltage sequences ($U(t) = U^{+}(t) + U^{-}(t) + U^{0}(t)$, in which $U^{+}(t)$, $U^{-}(t)$, and $U^{0}(t)$ are positive, negative, and zero sequences of $U(t)$), respectively. As a result, using the Optimal Solution theory (OS theory), the source-end current can be rewritten as

$$\begin{cases} \begin{cases} i_g(t) = i_g^{+}(t) + i_g^{-}(t) + i_g^{0}(t) \\ i_g^{+}(t) = \lambda \cdot U^{+}(t) \\ i_g^{-}(t) = \lambda \cdot U^{-}(t) \\ i_g^{0}(t) = \lambda \cdot U^{0}(t) \\ \lambda = \dfrac{\overline{P}(t)}{U(t).U(t)} \end{cases} \\ \begin{cases} i_g(t) = \dfrac{\overline{P}_g(t)}{U(t).U(t)} U(t) \\ i_C(t) = i_{Load}(t) - \dfrac{\overline{P_g}(t)}{U(t).U(t)} U(t) \end{cases} \end{cases} \tag{6}$$

where $i_g(t)$, $i_c(t)$, $i_{Load}(t)$, λ, and $P_g(t)$ are the source current, the compensation current, the current that must be compensated, the instantaneous power factor, and the instantaneous power, in the same order. In Eq. (6), $U(t)$ is the source of distortion due to the fact that it is non-sinusoidal.

Distorted current will be injected by the SAF compensation algorithm. For this reason, the compensation algorithm derived from the GTIP under the two asymmetric and distorted three-phase load-terminal voltages supplies unacceptable outcomes. To tackle these problems, a solution is adopted on the basis of A-GTIP theory.

In other words, a non-sinusoidal load current in addition to $i^{+}(t)$ is composed of $i^{-}(t)$ and $i^{0}(t)$. Negative and zero sequences must be supplied with SAF and positive sequence with source ($i_g(t)$). But due to U (t) is non-sinusoidal and consist of positive, negative and zero sequences, based on equation (6), the calculate $i_c(t)$ (SAF injected currents) cannot remove total negative and zero sequences of $i_g(t)$. This means compensation is not optimal.

- One suggestion to overcome voltage asymmetry is to replace U (t) by $U^{+}(t)$ in Eq. (6). Hence, the new source-end currents and the SAF-injected currents are obtained as follows:
- The source-end currents remain purely sinusoidal, while $U^{+}(t)$ does not include any harmonic components. Apart from that, the non-sinusoidal $U^{+}(t)$ in the term $U^{+}(t)$ acts as the source of distortion. Therefore, the SAF compensation algorithm will inject a distorted current. The SAF new-injected current will cause sinusoidal source-end currents in four-wire systems as follows:

http://dx.doi.org/10.5772/intechopen.79685

$$i_g(t) = \frac{\overline{P}_g(t)}{U_1^+(t).U_1^+(t)} U_1^+(t)$$
$$i_C(t) = i_{Load}(t) - \frac{\overline{P}_g(t)}{U_1^+(t).U_1^+(t)} U_1^+(t) \tag{7}$$

In which $U_1^+(t)$ is the fundamental component of $U^+(t)$. The SAF-controller block diagram is shown in **Figure 2**.

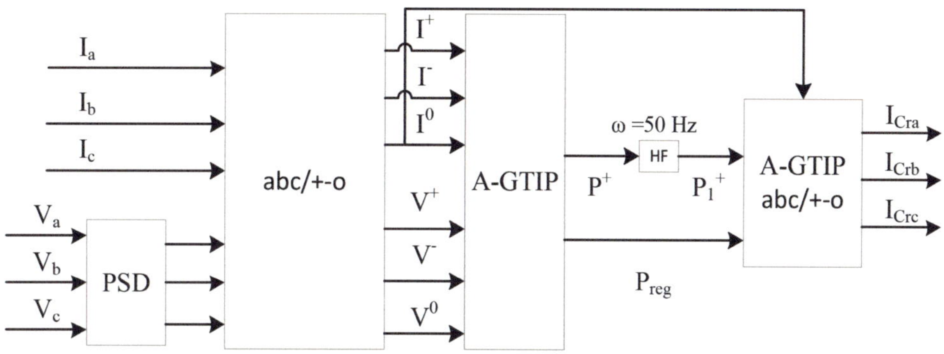

Figure 2. A SAF controller block diagram [1].

In this section, how the PV cells and SAF can be modeled has been explained; in the next section, how the grid-connected PV system and SAF can be related and the proposed block diagram will be presented.

3. SAF-PV system

Figure 3 illustrates the operating principle and current wave form I_{Load} at the load. The PV system is modeled as two parallel current sources. The first one, I_{PV}, is proportional to the

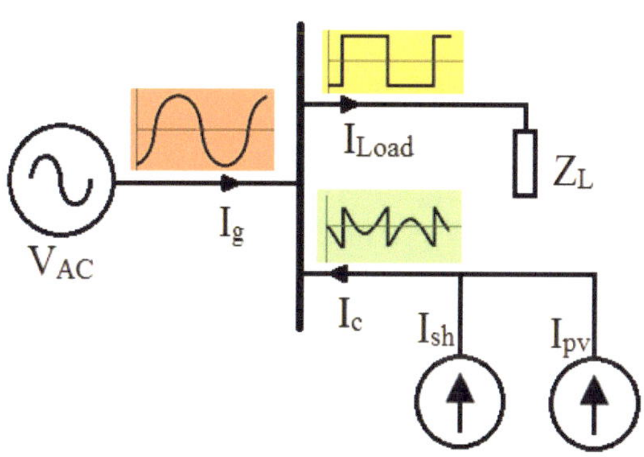

Figure 3. The operating principle of a SAF-PV system.

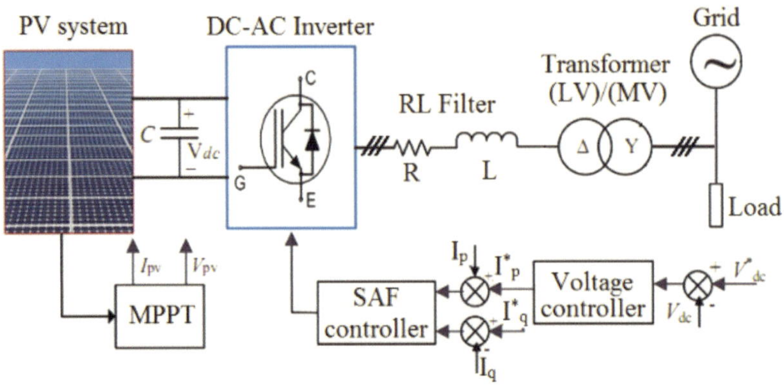

Figure 4. A schematic diagram of the single-stage SAF-PV system.

maximum power available from the PV cells, and its frequency and phase are equal to those of the voltage of the mains. The second current source supplies a wave form, which is equal to the total amount of the harmonics drawn by the load. The current supplied by the mains, I_g, is a purely sinusoidal wave, and the PV system reacts as an active filter to some of the active power and harmonic currents drawn by the load.

The schematic diagram of **Figure 4** shows the power stage of the grid-connected single-stage PV system. It includes the PV array, maximum power point tracking (MPPT) technique, which was used to extract the maximum available power from the PV array, and the DC-link capacitors that connect to the output terminal of the PV array. In addition, a three-phase VSI with its control is based on SAF, RL filter, which is connected to the low voltage AC grid, and a step-up transformer-connected distribution side of the grid.

The design of SAF incorporated with PV system can be decomposed into two issues: (1) maximum power point tracking (MPPT) of the PV system and (2) a control strategy of the voltage of the DC link (V_{DC}) common between SAF and PV systems.

3.1. MPPT

The relatively higher cost required for generating this type of energy in comparison with the energy produced by conventional power generation systems or other renewable sources such as wind power is known as the main disadvantage of the PV systems. Therefore, the optimal operation of the PV systems is critical and achieved by maximizing the efficiency of power delivered to the output by tracking the maximum power point. The PV system is connected to the grid via DC-DC converters. MPPT in PV systems is achieved by applying a control signal to the converters and regulating the PV terminal voltage (or current).

MPPT not only enables an increase in the power delivered from the PV module to the load but also enhances the operating lifetime of the PV system [4]. The solar cell maximum output power at the appropriate operating point and a given cell efficiency depend on the radiation intensity, ambient temperature, and load impedance. It is essential to ensure the efficient operation of the solar cell array that there is a single operating point in which through

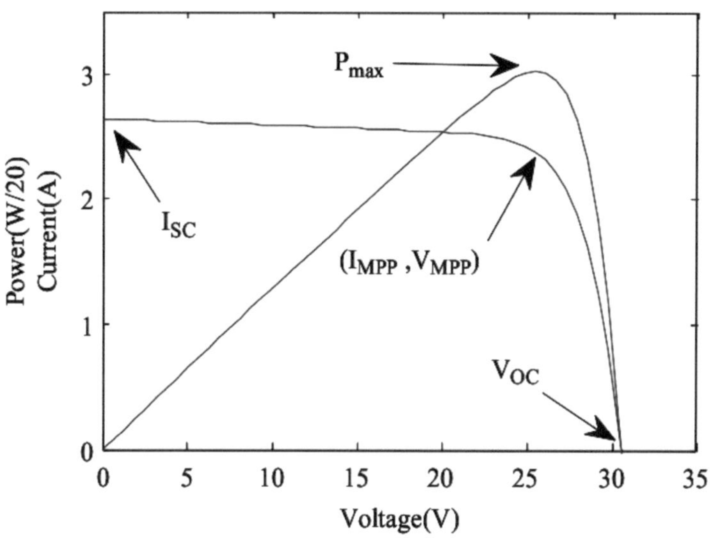

Figure 5. I-V and P-V characteristics of solar cell.

variations in radiation intensity and temperature, the maximum power point tracking is achieved (**Figure 5**). The problematic aspect of MPPT is that PV arrays automatically produce a maximum output power determined by PV output voltage or output current under a given temperature and irradiance. The maximum power attainment involves the adjustment of a load line under variations in temperature and irradiation level.

A wide variety of algorithms and methods have been proposed and implemented to attain MPP tracking [4–6], categorized namely offline methods, online methods, and hybrid methods. Offline methods are dependent on solar cell models, online methods do not specifically rely on the modeling of the solar cell behavior, and hybrid methods are a combination of the two abovementioned methods. On the other hand, the offline and online methods can also be referred to as the model-based and model-free methods, respectively.

3.1.1. *Offline methods*

In case of offline methods, it is generally required to know one or more of the solar panel values, such as the open-circuit voltage, V_{OC}, short-circuit current, I_{SC}, temperature, and irradiation. Voc and I_{SC} are two values that can be calculated based on measurement of the solar irradiance and temperature or be measured by applying an open circuit or a short circuit to the PV system. While the accuracy of the calculated values is limited by the accuracy of the PV characteristic provided by the manufacturer's specifications, the latter approach does not involve the load interruption necessary for measuring the V_{OC} and I_{SC}. The two values, V_{OC} and I_{SC}, are employed to generate the control signal, which is necessary for driving the solar cell to its maximum power point (MPP). In the course of the tracking of maximum power point operation, the abovementioned control signal remains constant if ambient conditions can be regarded as fixed and there are no attempts to regulate the output power of the PV system.

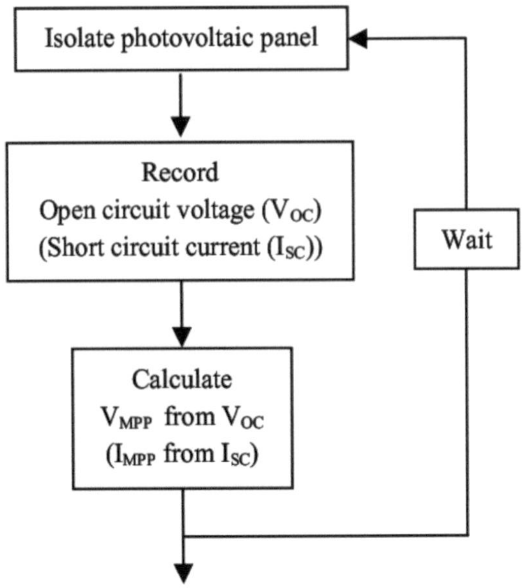

Figure 6. A flowchart of open-circuit voltage (short-current circuit) method [4].

The offline methods known as open-circuit voltage method (OCV), short-circuit current method (SCC), as well as the MPPT method based on artificial intelligence (AI) take variables such as temperature and irradiation as input and calculate the MPP. I_{SC} and V_{OC} can be measured or calculated based on mathematical models provided by the manufacturer or based on experimental data, which reflect the dependence on temperature or irradiance. AI models extract learning-based models using the known relationship between OCV or SCC and temperature and/or irradiance. A flowchart of these methods, which are the simplest offline methods, is depicted in **Figure 6**.

Both of OCV and SCC methods cannot deliver the maximum output power to the load because of two reasons. First, load interruption occurs during the measurement of I_{SC} or V_{OC}, and the second reason is that MPP can never be tracked quite exactly using these methods in the first place as suggested by approximately linear relationship between the open-circuit voltage V_{OC} and V_{MPP} or I_{SC} and I_{MPP}.

These two methods cannot be categorized as "true seeking" MPP methods; however, the simplicity of these algorithms and the ease with which they can be implemented make them suitable for use as part of novel hybrid methods [7, 8].

Artificial intelligence (AI) techniques, as other offline methods, are becoming popular as alternative approaches to conventional techniques or as components of integrated systems. They have been used to solve complicated practical problems in various areas. In [9], only the applications of artificial neural networks (ANNs) and fuzzy logic (FL) are discussed, while AI techniques consist of several disciplines.

The ANN-based method advantage lies with the fact that the trained neural network can provide a sufficiently accurate MPPT without requiring extensive knowledge about the PV parameters. Most of PV arrays exhibit different output characteristics; however, it must be mentioned that an ANN has to be specifically trained for the PV array with which it will be used. The time-varying characteristics of a PV array imply that the neural network has periodically trained to be guaranteed to track MPP accurately. Implement training periodically needs collecting data, which is a time-consuming process.

Fuzzy logic controllers take full advantages of the following: the ability to work with imprecise inputs, the requirement shortage of an accurate mathematical model, the ability to handle nonlinearity, and fast convergence. However, the approximation of achieving learning ability and accuracy depends on the fuzzy level number and the membership functions form. In most fuzzy systems, there is a connection between membership function, fuzzification and defuzzification, as well as the antecedent and the consequent fuzzy rules that are determined through trial and error, which can take a long time to perform.

3.1.2. Online methods

In case of online methods, the control signals are usually generated by using the instantaneous values of the PV output voltage or current. The control signal is applied to the PV system along with a small methodical and premeditated perturbation in voltage or current or duty cycle (control signal), and the resulting output power is determined. By analyzing perturbation response on the output power of a PV panel, the direction in which the control signal changes (decrease or increase) is determined. Hence, in contrast to the offline methods, the control signal can no longer be regarded as constant when a perturbation is applied. Therefore, the maximum output power tracking involves some oscillations around the optimum value.

In online methods, also known as model-free methods, control signals are usually generated by the instantaneous values of the PV output voltage or current. The more known online methods are Perturbation and Observation method (P&O), Extremum Seeking Control method (ESC), and the Incremental Conductance method (IncCond).

P&O method is considered by a number of researchers due to the fact that it is one of the simplest online methods [10]. P&O can be implemented by applying perturbations to the reference voltage or the reference current signal of the solar panel. **Figure 7** depicts this method's flowchart, which is also known as the "hill climbing method," where "X" is the reference signal. In the algorithm, taking the reference signal, X, as the voltage, (i.e., X = V), the goal will involve pushing the reference voltage signal toward V_{MPP}, thereby causing the instantaneous voltage to track the V_{MPP}. As a consequence, the output power will approach the maximum power point. With this end in view, a small but constant perturbation is applied to the solar panel voltage.

A systematic ECS methodology supported by rigorous theories such as averaging and singular perturbation was recently presented. This real-time optimization methodology involves a nonlinear dynamic system with an adaptive feedback. This ESC method has been successfully applied in PV systems in order to track MPP [11]. With the self-optimizing extremum algorithm as the MPPT controller, the control objective is for the PV system operating point to rapidly trace the MPPs subject to uncertainties and disturbances from the PV panel and the external load.

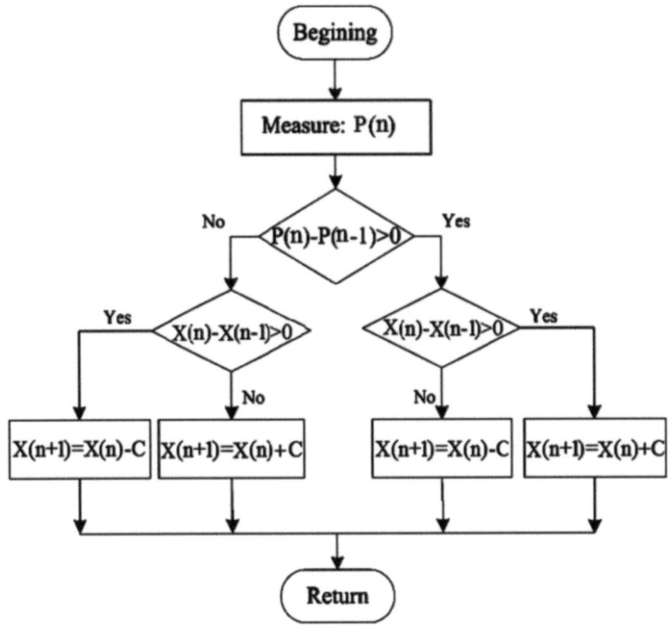

Figure 7. Perturbation and observation algorithm [4].

A small sinusoidal current represented by $\Delta I = a\sin(wt)$ is supposed and added to the reference current (I_{ref}) as a perturbation. This leads to making a ripple on the power (ΔP), whose phase and amplitude are dependent on the relative location of the operating point relative to the MPP. The sinusoidal current perturbation will be added to the reference current and applied to the PV system, as it is clear in **Figure 8**. If the resulting ripple in the current is in phase with the output power ripple, the output power will fall to the left of MPP, and the reference current will be less than I_{MPP}; therefore, the controller will increase the reference current.

On the other hand, if the ripple in the current is not in the same phase with that in the output power, the output power will drop down to the right of MPP, and the reference current will exceed the I_{MPP}. As a result, the controller will reduce the reference current until reaching MPP. The ripple power (ΔP) can be extracted by passing the output through a high-pass filter. Then, the ripple power is demodulated through multiplication by a sin ($wt- \varphi$) signal. The resulting signal, zeta, is either positive or negative depending on the position in the power output curve. After that, zeta is applied to an integrator in order to modify the value of I_{ref} to reach MPP. In the case of falling in the MPP operating point, the amplitude of the ripple will be negligible, and the output power ripple frequency will be twice as many as the current ripple.

There are two major advantages for ESC approach. The first is the optimization problem involving power maximization that is explicitly solved by using the dynamic adaptation-based feedback control law for a sinusoidal perturbation. Hence, attainable MPP is guaranteed when the control algorithm is convergent. The disadvantage of the ESC method lies in the complexity associated with its implementation as well as the necessity to evaluate signals of relatively low amplitude.

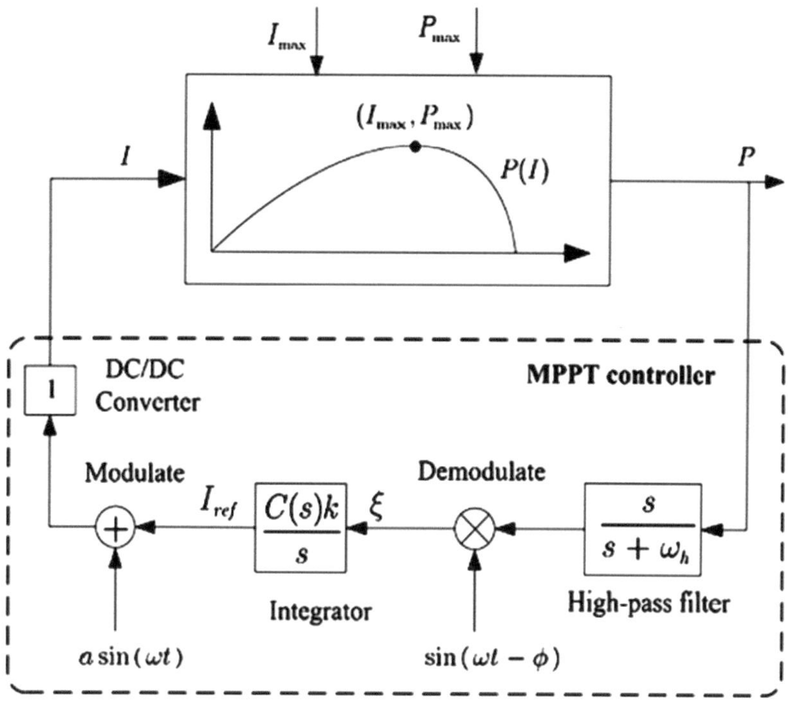

Figure 8. An MPPT controller scheme for the PV system [4].

The method employs the slope of the PV array power characteristics to track MPP that is known as incremental conductance (IncCond) method [12]. In this method, the curve slope of the PV array power is indicated. Based on this, it is zero at the MPP, positive for the output power values, which are smaller than MPP, and negative for values of the output power greater than MPP. The amount of the increment or decrement indicates the MPP tracking speed. An incremental increase may lead to fast tracking, but due to some oscillations around the MPP, the system may not exactly operate at the MPP. That is to say, the usage of IncCond method would be a sort of trade-off between convergence speed and the likelihood of causing oscillations in the MPP.

The main advantage of this algorithm is that it offers an effective solution under rapidly changing atmospheric conditions. The main drawback associated with the IncCond method is that it requires a complex control circuitry.

3.1.3. Hybrid methods

Hybrid MPP methods, a combination of the offline and online methods, are based on tracking of the MPP which are performed in two steps of estimation and exact regulation of MPP. The MPP estimation step relies on offline methods to place the set point close to MPP. The next step, regarded as a fine-tuning step, relying on online methods, attempts to reach the MPP actual value. As expected, the hybrid methods are more efficient to track MPP. In hybrid methods, the associated control signal has two parts, which are generated based on a separate algorithmic loop.

The first one is determined according to one of the simplified offline methods as a constant value, which depends on the given atmospheric conditions of the PV panel and represents the fixed steady-state value. In this first part, control signal is not intended to track the MPP accurately, while it is required for a fast response to the environmental variations. This part can be generated using one of the previous offline methods or simplifications based on the relationship between output power characteristics and ambient.

The second part obtained based on an online method involving steady-state searches represents attempts to follow MPP exactly. In contrast to the previous part, this second part attempts to reduce the error in steady state and does not require a fast response to the environmental variations. The algorithm, which is provided in **Figure 9**, represents a general description of the hybrid method. As pointed earlier, the first part of the control signal is generated using an offline method through the set-point calculation loop. By employing an online method via the fine-tuning loop, the second part is obtained.

A hybrid method, which has two loops, is presented in [13]. In the first estimation loop based on the open-circuit voltage at a constant temperature, MPP is approximated. In the second precise loop, the P&O method is applied to seek the exact amount of the maximum output power. The transient and steady-state responses are improved by maintaining the small amount of amplitude and frequency of perturbation. The authors in [6] proposed a hybrid approach in which an offline method is used to bring the operating point of the PV array near to the MPP. After that, an online IncCond method is used to track the MPP with high accuracy. Through proper control of the power converter, the initial operating point is set to match a

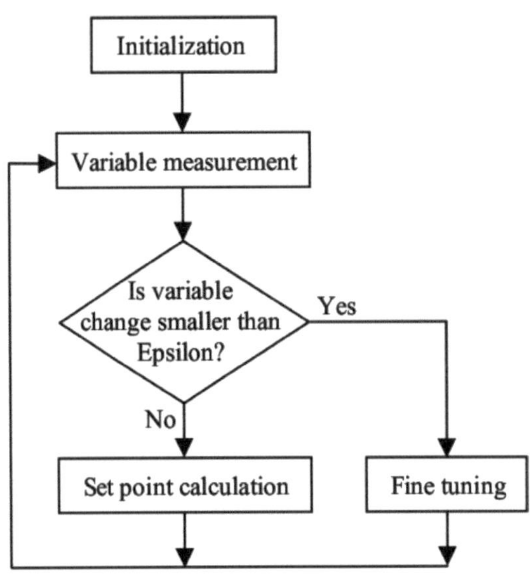

Figure 9. General algorithm of hybrid methods [6].

load resistance proportional to the V_{OC}/I_{SC} ratio associated with the PV array. In this hybrid method, the real MPP tracking is able to ensure that multiple local maxima are presented.

Implementing variable size perturbations by fuzzy logic is a matter of discussion in [14], in the context of achieving improved transient and steady-state responses. The converter duty cycle is adjusted to move the operating point toward the MPP region as soon as possible, thereby improving the response of transient state. A modified P&O algorithm that works based on fuzzy logic and optimized for small variations around the MPP is used simultaneously when MPP region is reached. This method decreases oscillations and increases power produced under the steady-state conditions. Every loop in these chapters applies the P&O approach using perturbations of different amplitude. Here, fuzzy logic decides which loop should be implemented. The peak current control as well as the abovementioned method can result in improving the transient response and decreasing the power loss under steady-state conditions, simultaneously [14].

3.2. DC-link voltage control

One significant feature of the PV-SAF is that typically a DC capacitor is connected between the voltage converter of a PV system and an SAF inverter, rather than a DC source. Because neither PV nor SAF is lossless, a special DC-link voltage controller is required to maintain the DC capacitor average voltage at a constant level. In the PV-SAF, the shunt active filter is usually responsible for this voltage regulation. In the steady state, the average DC-link voltage is maintained at a certain preset level, but during the transient, this is not the case. Such a transient can occur when a change occurred in the output power of a PV plant or a load is either connected or disconnected to/from the SAF. Since it takes a finite-time interval to calculate the new reference current, the shunt compensator cannot immediately respond to the load change. In addition to this, some settling time is required to stabilize the controlled parameter around its reference. Consequently, after a PV output power or a load changing instant, there exists some transient period during which the average voltage across the DC capacitor deviates from its reference value.

Figure 10 shows a strategy of DC-link voltage control in an algorithm, which has two main modes. The first mode would be when the PV produces the power that is ($P_{PV} > P_{min}$) delivered to the network through the SAF, while the second mode is when the threshold P_{min} is more than the power generated by the PV. It needs to be noted that in the two modes, V_{DC} must be greater than V_{min} to have a satisfactory operation of the SAF.

The PV-SAF grid power flow algorithm is illustrated in **Figure 11**. The SAF is like a load that varies when its consumed power is changed with the power generated by PV and the V_{DC}. In the first mode, a boost converter is applied to deliver the generated power of the PV to the DC link. As stated in the control strategy, when threshold voltage ($V_{DC\text{-}min}$) is greater than V_{DC}, the PV power is fed to the DC link in order to maintain it in an acceptable range.

In this situation, the power delivered to the grid through SAF (P_{sh}) is zero; in other words, the power of the PV is solely dedicated to charge the capacitor of the DC link. As V_{DC} exceeded the $V_{DC\text{-}min}$, in order to charge the capacitor, a portion of the P_{PV} is used, and in that case, the rest

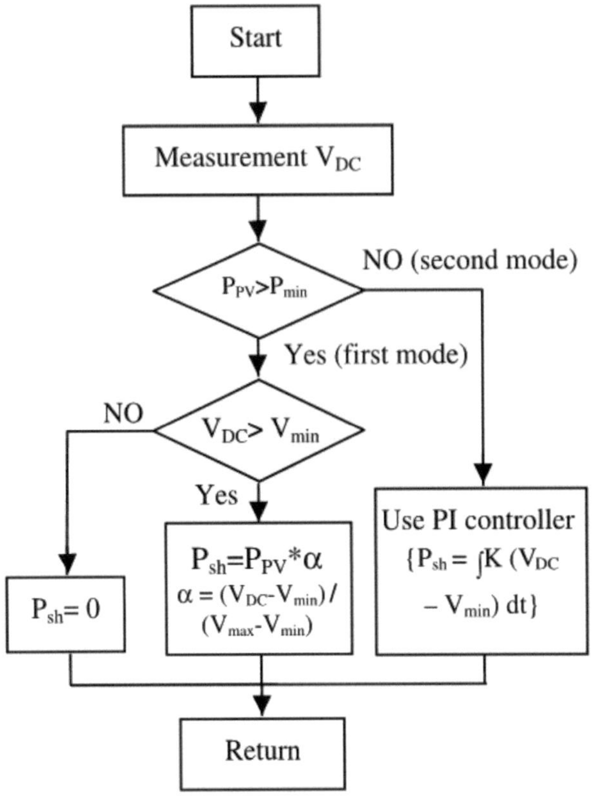

Figure 10. The overall algorithm of the DC-link voltage control [1].

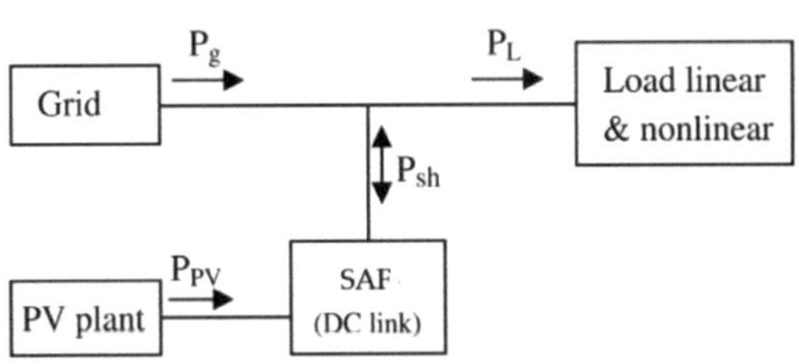

Figure 11. An algorithm of power flow of the PV-SAF [1].

will be delivered to the grid through the parallel part of the SAF. When V_{DC} reaches its maximum and allowable voltage, that is $V_{DC\text{-}max}$, the power of the PV is all fed to the grid.

In the situation that V_{DC} is between $V_{DC\text{-}min}$ and $V_{DC\text{-}max}$ there would be a linear relationship between the power used to charge the capacitor and the $V_{DC\text{-}max}-V_{DC}$ ($P_{sh} = \alpha\ P_{PV}$ and $\alpha = (V_{DC}-V_{min})/(V_{max}-V_{min})$). As a result, some of the load power has been supplied with the PV plant when V_{DC} is greater than $V_{DC\text{-}min}$. Following Kirchhoff's circuit laws, as seen in Eq. (8), the load current (I_{Load}) is equal to the sum of the grid current (Ig) and the SAF current (Ish). Also, SAF current (Ish) is equal to the sum of the PV power plant (I_{PVPP}) and compensated current (I_C). Here, Eq. (7) can be rewritten to Eq. (9) in order to control the shunt active filter to deliver power of the solar power plant.

$$\begin{aligned} &i_g(t) = i_{Load}(t) - i_{sh}(t) \\ &i_{sh} = i_{pvpp} + i_C \\ &\rightarrow P_g(t) = P_{Load}(t) - P_{sh}(t) \\ &i_{PVPP}(t) = \frac{\overline{P}_{sh}(t)}{U_1^+(t).U_1^+(t)} U_1^+(t) \end{aligned} \tag{8}$$

So, we have

$$\begin{aligned} &i_g(t) = \frac{\overline{P}_g(t)}{U_1^+(t).U_1^+(t)} U_1^+(t) \\ &i_C(t) = i_{Load}(t) - \frac{\overline{P}_g(t) - \overline{P}_{sh}(t)}{U_1^+(t).U_1^+(t)} U_1^+(t) \end{aligned} \tag{9}$$

As addressed in Eq. (9), in addition to providing all negative and zero components of the nonlinear load, SAF will partially provide the positive current component of the nonlinear load, which will result in a lower source current. In the case of a voltage sag, the stored energy in the DC link is fed to the grid through the series part of the SAF. This injected energy enhances the power quality but also causes a decrease in the voltage of the DC link, which will be compensated by the energy given by the PV. In a situation where the PV does not generate power (e.g., $P_{pv} < P_{min}$), the demanded energy to charge the DC link will be provided by the grid.

4. Simulation and discussion

The analysis of the three phases has been done in MATLAB/ SIMULINK environment. The system has a three-phase AC source of 230 V at 50 Hz which is represented as an ideal, balanced, delta, three-phase voltage source, feeding a three-phase nonlinear load (75KVA). The maximum generated power of the PV system is 60 KW.

PV-SAF is utilized for the improvement of power quality in which parallel-connected inverter is operated to perform line current harmonics elimination and reactive power compensation. To study the performance of the proposed algorithm, the results are presented for PV-SAF, where the solar power plant consists of several series-parallel solar panels, which are connected to a boost converter, and delivers the generated power of the solar power plant to the DC link. This power will be delivered by the active filter to a nonlinear load with TDH of more than 40% through a three-phase AC grid with a frequency of 50 Hz and a voltage of 230 V.

Figure 12 shows the voltage and power of solar power plant for MPPT methods of [1, 6]. For both methods of MPPT, an estimation of the operation point is used to set the operation point to a fairly close point near the MPP, and then by means of a fine-tuning loop, MPP will be reached. In the method of [6], the voltage of the MPP is approximated above the actual V_{MPP}

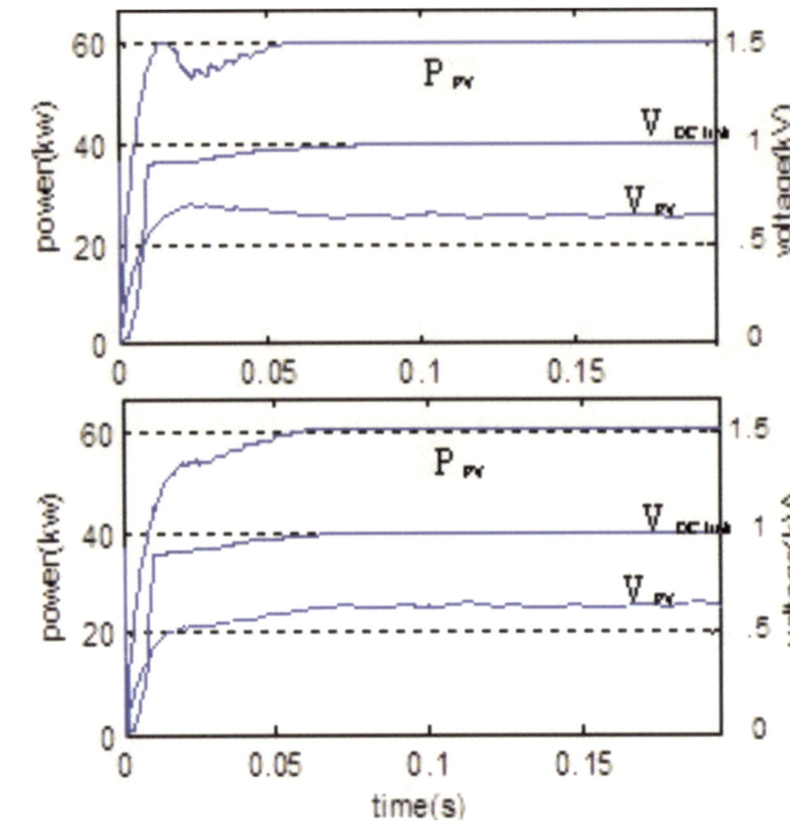

Figure 12. Voltage and power of the PV using MPPT methods of (a) [6] and (b) [1].

operation point consequently, by an increase of V_{DC} from 850 V to 1 KV; the operation point is deviated from MPP and then returns after several milliseconds. It should be noted that the slope of the P-V characteristic curve is greater for voltages more than V_{MPP} in comparison to smaller voltages, meaning that a small change in voltage leads to a great change in the power of the solar panel. It can be seen that in the method of [1], the power smoothly increases until it reaches the MPP. This is because of the independency of the control signal to V_{DC}.

Figure 13, I_{Load}, illustrates the simulated grid current without PV-SAF operation. It is obvious that the grid current is non-sinusoidal and consists of the 50-Hz fundamental component along with lower-order harmonics like the third harmonic (150 Hz), fifth harmonic (250 Hz), seventh harmonic (350 Hz), and so on. **Figure 13**, I_g, shows the grid current with PV-SAF operation. It is clear that the harmonic currents of nonlinear load are almost compensated with the PV-SAF operation so that the line current is nearly a sinusoidal wave. The total harmonic distortion of line current is lower than 4%. **Figure 13**, I_{sh}, shows the injected current of the shunt part of the PV-SAF to compensate the current harmonics of the load so that the grid current can be sinusoidal. Meanwhile, it delivers the power of the PV to the load.

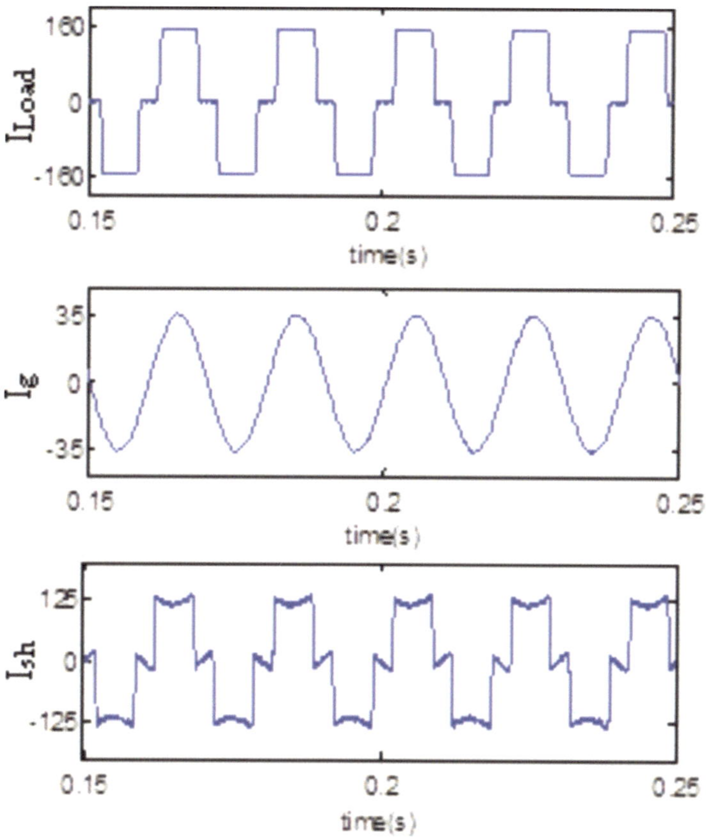

Figure 13. Grid current without PV-SAF (I_{Load}), grid current with PV-SAF (I_g), and the current of part shunt PV-SAF (I_{sh}).

5. Conclusions

In this chapter combined with power conditioner and renewable energy, SAF-PV system has been explained for the optimal designing of PV grid connected. Meanwhile, considering the Advanced Generalized Theory of Instantaneous Power (A-GTIP) algorithm, the SAF-PV system leads to suppress grid-end current harmonics caused by the distorted unbalanced load-terminal voltages. Hence, the grid-end currents could remain purely sinusoidal. Also, PV power is injected to the grid via active filter converter and MV/HV transformer. It means that by using the SAF-PV system, there will be capital investment savings since one less converter and MV/HV transformer will be used in comparison with separated SAF and PV systems. In this chapter, different maximum power point tracking (MPPT) algorithms have also been reviewed which can serve as a guide for the selection of the appropriate MPPT method for specific PV system applications. Various simulation results verify the performance of the combined PV-SAF.

Author details

Ali Reaz Reisi[1]* and Ashkan Alidousti[2]

*Address all correspondence to: reisi.alireza@gmail.com

1 Electrical Engineering Department, Technical and Vocational University, Isfahan, Iran

2 Young Researchers and Elite Club, Shahrekord Branch, Islamic Azad University, Shahrekord, Iran

References

[1] Reisi AR, Moradi MH, Showkati H. Combined photovoltaic and unified power quality controller to improve power quality. Solar Energy. 2013;**88**:154-162

[2] Albuquerque FL, Moraes AJ, Guimara CG, et al. Photovoltaic solar system connected to the electric power grid operating as active power generator and reactive power compensator. Solar Energy. 2010;**84**:1310-1317

[3] Pashajavid E, Bina MT. Zero-sequence component and harmonic compensation in four-wire systems under non-ideal waveforms. Przegląd Elektrotechniczny. 2009;**85**:58-64

[4] Reisi AR, Moradi MH, Jamasb S. Classification and comparison of maximum power point tracking techniques for photovoltaic system: A review. Renewable and Sustainable Energy Reviews. 2013;**19**:433-443

[5] Esram T, Chapman PL. Comparison of photovoltaic array maximum power point tracking techniques. In: IEEE Transactions on Energy Conversion. 2007;**22**(2):439-449. DOI: 10.1109/TEC.2006.874230

[6] Moradi MH, Reisi AR. A hybrid maximum power point tracking method for photovoltaic systems. Solar Energy. 2011;**85**:2965-2976

[7] Zhang C, Zhao D, Wang J, et al. A novel two-mode MPPT method for photovoltaic power generation system. In: IEEE 6th International Conference on Power Electronics and Motion Control. 2009. Wuhan, China: IEEE; 2009. p. 2100-2102

[8] Yang C, Hsieh C, Feng F, Chen K. Highly efficient analog maximum power point tracking (AMPPT) in a photovoltaic system. In: IEEE Transactions on Circuits and Systems I: Regular Papers. July 2012;**59**(7):1546-1556. DOI: 10.1109/TCSI.2011.2177008

[9] Salah CB, Ouali M. Comparison of fuzzy logic and neural network in maximum power point tracker for PV systems. Electric Power Systems Research. 2011;**81**:43-50

[10] Abdelsalam AK, Massoud AM, Ahmed S, Enjeti PN. High-performance adaptive perturb and observe MPPT technique for photovoltaic-based micro grids. In: IEEE Transactions on Power Electronics. 2011;**26**(4):1010-1021. DOI: 10.1109/TPEL.2011.2106221

[11] Lei P, Li Y, Seem JE. Sequential ESC-based global MPPT control for photovoltaic array with variable shading. IEEE Transactions on Power Electronics. 2011;**2**(3):348-358

[12] Mei Q, Shan M, Liu L, Guerrero JM. A novel improved variable step-size incremental-resistance MPPT method for PV systems. In: IEEE Transactions on Industrial Electronics. 2011;**58**(6):2427-2434. DOI: 10.1109/TIE.2010.2064275

[13] NSD' Souza LAC, Liu LXJ. Comparative study of variable size perturbation and observation maximum power point trackers for PV systems. Electric Power Systems Research. 2010;**80**:296-305

[14] Kobayashi K, Takano I, Sawada Y. A study on a two stage maximum power point tracking control of a photovoltaic system under partially shaded insolation conditions. In: IEEE Power Engineering Society General Meeting. Vol. 4; 2003. pp. 2612-2617

Chapter 7

Experimental Study of Current-Voltage Characteristics for Fixed and Solar Tracking Photovoltaics Systems

Chukwuemeka Ikedi

Additional information is available at the end of the chapter

http://dx.doi.org/10.5772/intechopen.79710

Abstract

The efficiency of solar electric systems basically depends on the materials used in making the solar cells and regardless of the type of application: fixed or tracking photovoltaics (PV), the quality and quantity of power produced by PV systems depend on both the amount of solar radiation incident on the solar panels as well as the current and voltage characteristics of the load. This present work, which involves field installation of a fixed PV alongside an existing equivalent tracking PV, simultaneously monitored the current and voltage response of both systems to changing solar radiation and ambient temperatures. The comparative results of the study provide a framework for decision-making on the choice of either of the systems and have shown that in the UK, both systems have a relatively slow electrical response to sunrise while the performance of fixed PV systems approximates that of tracking PV systems at noon time.

Keywords: fixed PV, solar tracking PV, voltage-current (I-V) characteristics, maximum power, solar radiation

1. Introduction

Photovoltaics, otherwise called photoelectricity, is a compound word for photo that is light and voltaic that is electric (from Volta the inventor). It is simply the conversion or generation of electricity from light. Jacques Bequerel, a French physicist discovered in the 1890s, that certain materials produce electric current when exposed to light. This phenomenon is called photoelectric effect and forms the basis for the science and technology of photovoltaics.

Photovoltaics (PV) technology has been in existence for more than 50 years now [1] with various innovative applications. For instance, the Swiss solar aircraft, "Solar Impulse 2", achieved the longest non-stop solo flight in history making the first solar-powered aerial circumnavigation in 2015. The wide range of fixed PV applications include—lighting (e.g. for buildings, streets, traffic signals and navigation), transportation (e.g. solar-powered vehicles, boats, ships), telecommunications, astronomy and space electrical power supplies.

Other applications include roadside emergency telephones, parking ticket machines, remote sensing and cathodic protection of oil and brewery pipe lines.

This chapter presents an overview of basic photovoltaics materials and components and most significantly, investigates and analyses the electrical characteristics of two types of installed PV systems namely: fixed and solar tracking PV simultaneously, under varying solar radiation and temperature conditions in the UK.

By deducing and comparing the maximum power for each of the systems at different points in time, interesting observations were made which led to vital conclusions regarding the relative choice of either of the systems with respect to their respective maximum power performances and cost under similar applications and conditions.

2. Photovoltaic materials and solar panels

In order to appreciate any solar electric or PV system and applications, whether experimental as in this present study or otherwise, a brief overview of the background becomes inevitable.

The basic component of a PV module or panel is the solar cell. Although recent research and innovations have identified some other materials, extensive literature reviews in this study has identified silicon as the key material used in making the solar cells for most PV panels and unless broken or exposed to harmful elements, they could last for a period of more than 20 years and usually protected behind transparent glass materials. Three major configurations of silicon photovoltaic materials were identified in use for solar panels, namely: monocrystalline, polycrystalline and amorphous cells.

Monocrystalline cells: These were found to be the first commercially developed solar cells. They are cut from single crystals of silicon and have an efficiency of about 11–16%. They are chemically stable [2]. They have negligible defects and impurities and are usually grown from a sophisticated but expensive process, known as Czochralski process.

Polycrystalline cells: Cut from many silicon crystals, they have a single colour tone, multiple patterns and an efficiency of about 9–13%. They are cheaper in production than monocrystalline types, but less efficient, because of their light-generated charge recombination effect [3].

Amorphous cells: These are silicon cells in non-crystalline form, usually used in thin film technology. They are cheaper to produce but have a lower operating efficiency of about 3–6% [2]. They decay over time and are usually used in devices like watches, calculators and toys.

Apart from silicon, other crystalline materials are found to be in use for PV solar cells. One of such materials is the compound semiconductor, gallium arsenide (GaAs). PV cells made from GaAs are more expensive but more efficient than silicon cells. They withstand high temperatures and are therefore used in concentrating PV systems and also for applications that demand very high efficiency, irrespective of cost, like in space operations.

Organic materials are now available for use in solar cells. Most organic photovoltaic cells are made from polymers and compared to silicon solar cells, polymer solar cells are lightweight and cheap to fabricate and have safer environmental impact. The capability to be transparent has made polymer solar cells useful in applications like windows, glass walls and skylight roofing devices. One downside with organic solar cells is that they produce a relatively low level of the efficiency compared to silicon materials [4].

Presently, research is advancing towards the development of more efficient and cheaper materials for generating photo electricity. A new approach, using millimetre-sized polycrystalline silicon spheres on thin sheet aluminium foils have been developed by Texas Instruments, US for PV cells.

It is important to note that each silicon solar cell usually has a voltage output of 0.5 V, and when a collection of such cells are electrically connected for the purpose of meeting certain specified load requirements, it is referred to as a PV module or panel. Furthermore, an electrical combination of two or more PV modules or panels to achieve a specific voltage and current as required by a given load or appliance in a particular application is referred to as PV arrays. The individual modules could be either similar or dissimilar and can be connected either in series or in parallel, unlike the case of cells in the module. The systems installed and used in this experimental study consists of an array of two similar solar PV modules electrically connected in series for either of the applications: fixed and tracking. In general, arrays provide increased power output.

The other components of a PV system include the battery, the charge controller and the inverter and when all connected together becomes referred to as photovoltaics generator.

2.1. Fixed and tracking PV

This experimental study simultaneously monitored a fixed and a tracking PV system. It becomes important therefore to provide a brief explanation of both systems. The Earth moves round the Sun in an elliptical orbit; in a counter clockwise direction on an imaginary line called its axis, tilted with respect to the plane of its orbit at an angle of about 23.4°. Due to this movement of the Earth around the Sun and the consequent effect on solar radiation, some PV systems are designed to track the Sun's movement and hence maximise solar incidence on the modules/arrays by maintaining an optimum orientation between the Sun and the solar panels. Such systems are referred to tracking PV. The complex and usually delicate operations involved in tracking PV systems has meant that most PV applications are of the fixed category resulting in benefits of simplicity, least cost and convenience of operation.

On the other hand, fixed PV systems are defined as such because the solar modules or arrays are permanently fixed at a particular angle towards the Sun, with the aim of maximising solar capture. Fixed systems can be installed either as pole mounted, ground mounted or roof mounted systems.

Pole and ground mounted PV's as in this study are usually installed remote from building envelopes, while other types of PV systems are either installed on structured framework on the roofs of buildings or integrated with the building envelope in such a way that it is referred to as building integrated photovoltaic (BIPV). These involve the integration of the PV modules into parts of the fabric of a building as roof tiles, asphalt shingles, facade materials or shading elements. Used in this way, the integrated PV modules replace conventional building envelope materials thereby benefiting from capital cost reduction and hence improved payback period and life cycle cost.

3. Research methodology

As a preliminary step, an extensive review of previous works was carried out prior to this present study. Most work and research carried out earlier on PV materials were found to be mainly on cell characterisation and development [5–7], etc. Shivakumar et al. carried out a test on interface adhesion strength in multilayered structures; Dauskardt et al. examined the mechanisms of debonding in photovoltaic back sheets; Budiman et al. applied Synchrotron X-ray on c-Si Solar PV cells for micro-diffraction analysis, and these are to mention a few. Few researches conducted on solar tracking PV suggest the average experimental gain of tracking PV systems over the fixed types to be about 25% [8–11].

However, when the use or application of fixed and tracking PV systems is considered at different seasons of the year putting into consideration, obtainable costs of maintenance, a controversy begins to arise which seems to question the credibility of the claimed gain of the tracking PV system over the fixed option. Comparative study on specific aspects of the systems such as power outputs gives a clearer understanding of the respective performances.

Further reviews were carried out on current voltage (I-V) characteristics of photovoltaics materials [12–14]. James et al. showed I-V characteristics with reverse bias slopes to be due to wavelengths of light below semiconductor band gap, while Schottky I-V characteristics were due to wavelengths of light above the semiconductor band gap; Zhang et al. proposed a method to predict I-V curves under different operating conditions, while Ibrahim investigated the response of crystalline silicon (Si) solar cell at different conditions of solar irradiance and showed possible performance defects.

This present study simultaneously monitors and compares the voltage–current response of fixed and solar tracking PV systems under the same varying conditions of solar radiation and ambient temperatures.

3.1. Identification and selection of research case study

As a further preliminary step to achieve a good experimental method, a case study was identified for use in the comparative analyses for the two systems, fixed and tracking PV, respectively. One key requirement for the selection of a case study in this study is that the fixed

http://dx.doi.org/10.5772/intechopen.79710

and tracking PV systems should be installed within the same location and at the closest possible vicinity to each other. This becomes necessary to ensure that the ambient temperatures around the two systems are significantly the same under the same solar insolation. Another key requirement or criteria for the selection of a case study is that the two systems must be of the same system specification and size.

The exact location used for the experimental study is the school of the built environment at the University of Nottingham, UK. The geographical and meteorological details of the location are as follows: Latitude 52.5° North, Longitude, Altitude 48 m and Azimuth 0° (true south).

For the photovoltaic systems, the fixed and the solar tracking PV consist of 2 PV modules tied together in serial connections, respectively. The PV module used for the installations is a BP 275F solar module with a nominal peak power (Pmax) of 75.00 W, maximum power voltage (Vmp) of 17.0 V and maximum power current (Isc) of 4.45 A. The extra features on the tracking system include the solar tracking sensor, made of monocrystalline cells and a 24 inch actuator motor jack for the tracking mechanism. The tracking PV system which was originally used to power a water fountain was already existing while the tracking PV was installed right beside the tracking system for the purpose of the comparative analyses in the study.

3.2. Experimental method

The experimental rig for the study consists of fixed and tracking PV systems, each made up of two BP 275 PV modules with the terminals in each system applied to the electrical circuit in **Figure 1**. The two systems were installed to have the same orientation, south facing at zero azimuths with module inclination of 52°s (which approximates the latitude of the location), having a nominal standard test condition (STC) open-circuit voltage of 21.4 V, short-circuit current of 4.75 A and peak power of 75 W.

STC is an abbreviation for the "standard test condition" by which PV modules are tested and calibrated which is insolation level of 1000 W/m^2, air mass of 1.5 AM and cell temperature of 25°C.

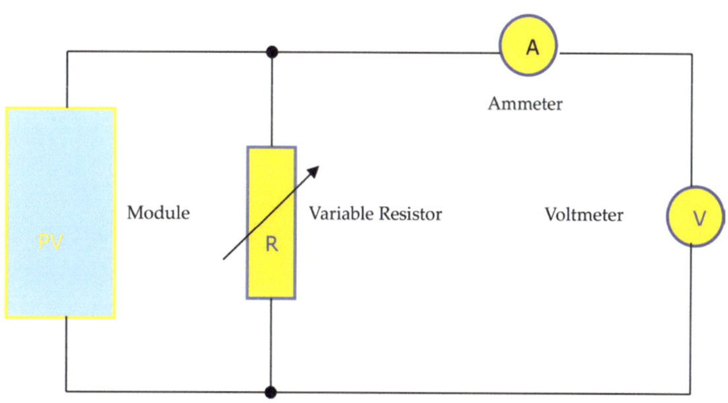

Figure 1. Layout of I-V electrical circuit.

While one of the systems remained fixed relative to the position of the Sun, the other (tracking) kept moving automatically with the aid of a solar tracking sensor and mechanical actuator jack to follow the changing positions of the Sun.

By using a potentiometer type of rheostat, the impedance in the circuitry (**Figure 1**) was varied, while the corresponding current and voltage at each point was monitored and recorded with the aid of the ammeter and the voltmeter.

This process was carried out every 1 h for 2 days between the solar window from 12.00 pm to 4.00 pm for the first day and 9.00–3.00 pm for the second day. One major problem encountered during the measurements was the dramatic change/drop in observed values in some cases due to sudden changes in insolation. This was because UK unlike some tropical locations has a very sloppy insolation gradient within the solar window such that each change in the insolation implies a big difference in the observed values.

4. Results and discussions of experimental work

From the maximum power and the I-V curves, it can be noticed that the gap between the curves for the fixed and tracking systems at each point in time oscillates from infinity towards zero and then towards infinity with noontime as a turning point from sunrise to sunset respectively.

This is because, around and within noontime, the fixed PV system sees the Sun at approximately a perpendicular position and at such point in time also, the tracker device in principle positions the tracking PV system at the same position, hence the difference in performance between the two systems becomes apparently cancelled and so the fixed system almost, approximates to the tracking system.

From noontime towards either sunrise or sunset, the effect of the tracking device on the tracking PV system becomes pronounced as the system becomes more resolved in orientation to the Sun relative to the fixed system.

From basic PV principles [15], the current flowing in the circuit above (**Figure 1**) at each point in time, can be given as

$$I = I_L - I_D(V) \tag{1}$$

assuming a linear superimposition of the photo and dark currents where the photocurrent

$$I_L = eA_C \int_{E_G}^{\infty} S(E)\left[1 - \rho(E, W) - \tau(E, W)\right] dE \tag{2}$$

and the dark current

$$I_D(V) = I_0^{\left[\exp\frac{eV}{(mkT-1)}\right]} \tag{3}$$

where $\alpha(E, W)$ is the spectral absorbance, $S(E)$ is the number of photons of energy E incident on the cell per unit area, A_C is the area of the illuminated cell, k is the Boltzmann's constant, T is the absolute temperature, and m is 1 at high voltages and 2 at low voltages.

From the graphs, (**Figures 2** and **4**), when the resistance is zero, the current in the circuit becomes the maximum (short-circuit current). At this point, the voltage $V = 0$ and from Eq. (1), the short-circuit current becomes

$$I_{SC} = I(V = 0) = I_L \quad (4)$$

Also from the same graphs, at open circuit, the current becomes zero while the voltage becomes the maximum (open-circuit voltage)V_{OC} and expressed as:

$$V_{OC} = \frac{mkT}{q\ln\left\{\frac{I_L}{I_0 + 1}\right\}} \quad (5)$$

For each of the cases, the area under the curve, which is the product of the voltage and the current, gives a measure of the power output.

At V_{OC} and I_{SC}, the power output becomes zero and maximum at a particular point between these points. This is the point at which the systems deliver the maximum power (maximum power point).

The values of the voltage and current at such points denote the maximum power voltage and the maximum power current V_M and $I_{M,}$ respectively.

The monitored and measured results for the different days are shown and described as follows:

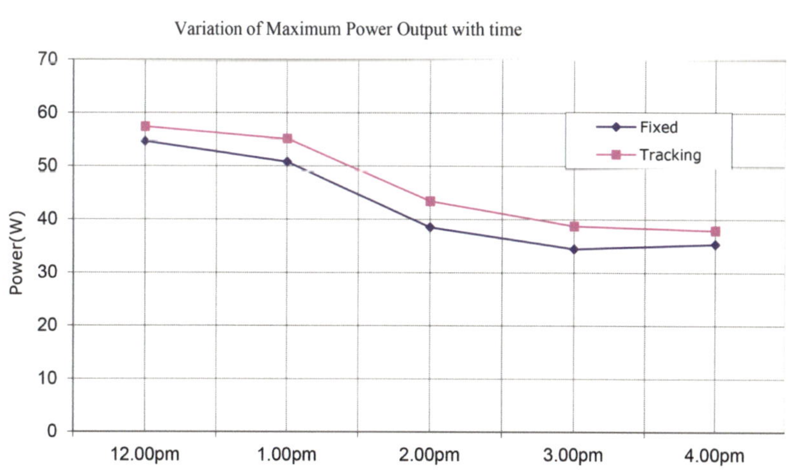

Figure 2. Timely maximum power output.

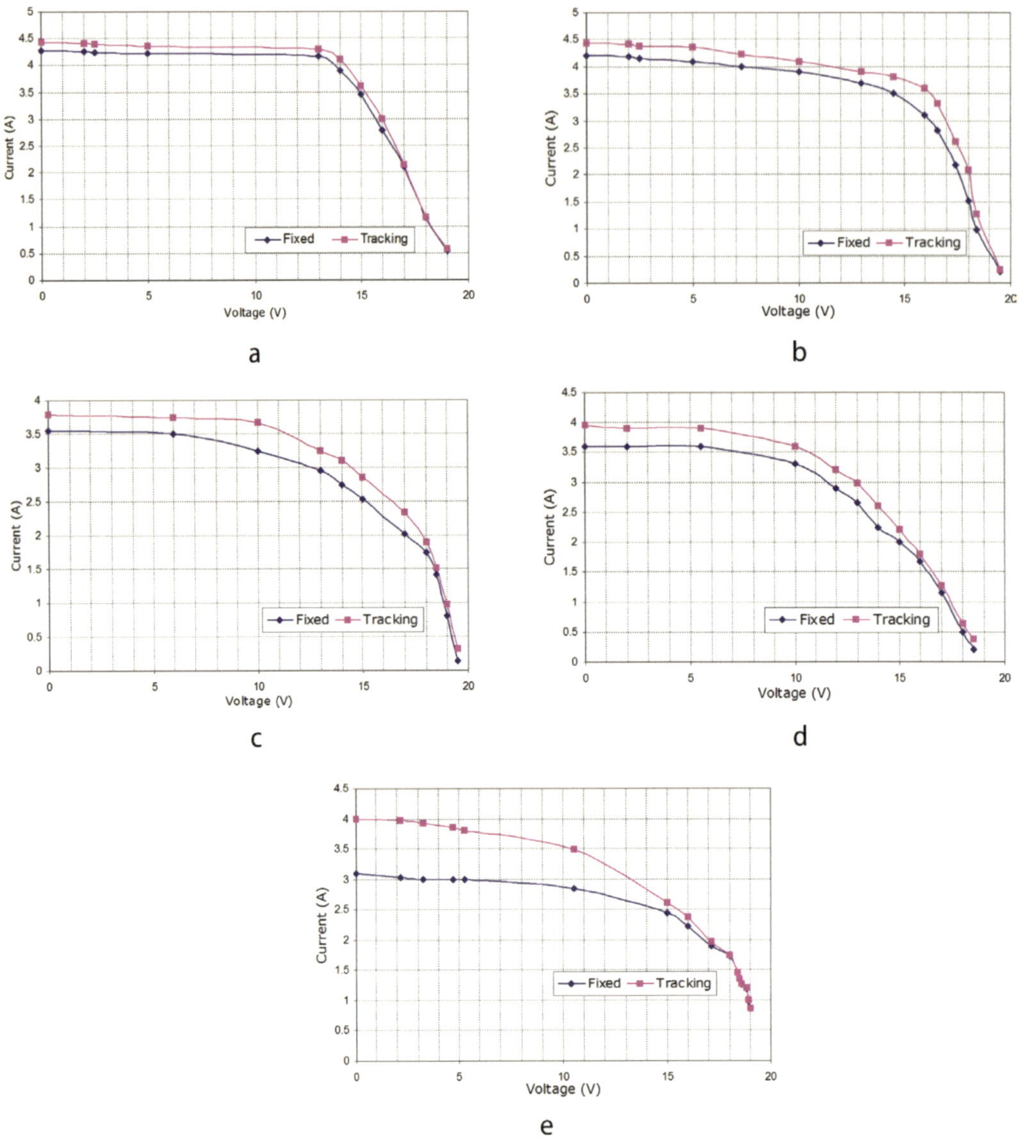

Figure 3. (a) I-V curves at 12.00 pm (hor rad. – 667.9 KWh/m^2. Ambient temp: 22.3°C), (b) I-V curves at 1.00 pm (hor rad. – 517.8 KWh/m^2. Ambient temp: 20.7°C), (c) I-V curves at 2.00 pm (hor rad. 616 KWh/m^2. Ambient temp: 22.7°C), (d) I-V curves at 3.00 pm (hor rad. 517.1 KWh/m^2. Ambient temp: - 21.2°C), (e) I-V curve at 4.00 pm (hor rad. 509.8 KWh/m^2. Ambient temp: 20.8°C).

4.1. Day 1

Table 1 contains the values of the short-circuit current and the open-circuit voltages for each of the observations and summarises the results of the measurements and observations carried out on day 1.

Below is the diagram comparing the maximum power output for the two systems (fixed and the tracking PV) for day 1.

http://dx.doi.org/10.5772/intechopen.79710

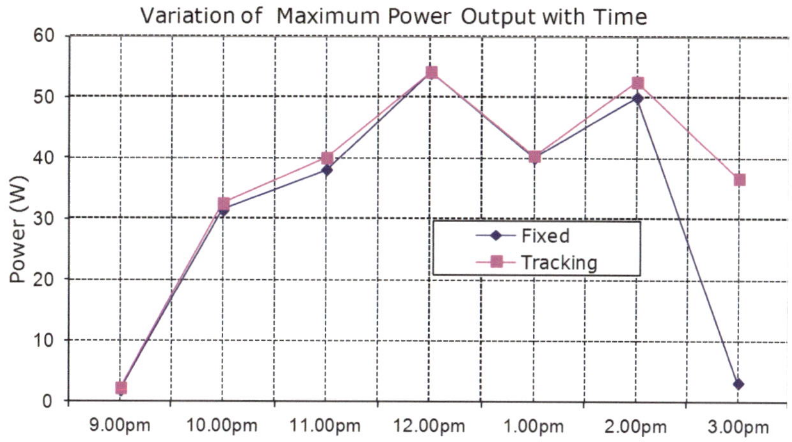

Figure 4. Timely maximum power output for day 2.

Time	Vmp (v)		Imp (A)		Pmp (W)		Voc (V)		Isc (A)	
	Fixed	Trk	Fixed	Trk	Fixed	Trk	Fixed	Trk	Fixed	Trk
12 pm	14.0	14.0	3.90	4.10	54.60	57.40	19.0	19.0	4.28	4.42
1 pm	3.50	3.80	3.70	3.80	50.75	55.10	19.5	19.5	4.20	4.43
2 pm	14.0	14.0	2.75	3.10	38.50	43.40	19.5	19.5	3.54	3.79
3 pm	13.0	13.0	2.65	2.98	34.45	38.74	18.5	18.5	3.60	3.95
4 pm	16.0	16.0	2.21	2.37	35.36	37.92	19.5	19.5	3.09	4.00
Average maximum power (W)					42.73	46.51				
Percentage power gain by tracking (%)						8.85				

Table 1. Measurements and observations for day 1.

It should be recalled from above that power variation approximately oscillates with noontime as the turning point. This is evident from the graphs above as the entire graph for the case of the respective date, represents only about one half of the daily solar window. As earlier pointed out, this was due to the poor climatic condition (Uneven radiation gradient) during the earlier part of the day before noontime.

Nevertheless, the picture of the entire cycle may be observed in the next investigation (**Figure 3**).

It can be concluded from the graphs that the average daily maximum power of the two systems increases from sunrise and peaks around noontime and then gradually decreases towards sunset with the tracking system maintaining a higher output at all times.

The response to sunrise and sunset generally depends on season and climatic conditions of a location [16], for instance, in temperate locations like some parts of Africa, PV panels would be readily responsive to the solar position as at 7.30 am where as in the UK during the time of this experimental work, the graphs of the two systems (fixed and tracking) were almost parallel to the voltage axis as at 9.00 am (**Figure 4a**).

Finally, the voltage-current characteristics was plotted and investigated under varying load conditions (resistance), solar insolation and ambient temperature.

The diagrams below (**Figure 3a–e**) show the I-V curves every 1 h from 12.00 pm to 4.00 pm for day 1.

One interesting thing to note in the above graphs (**Figure 3a–e**) is the change in the power margin between the two systems. Around noontime, the power margin tends very close to zero. The reason for this is explained in the third paragraph of Section 5.

On the other hand, it tends towards infinity around sunrise and sunset. It should also be noticed that at lower load impedance (maximum voltage), the two graphs in most cases begin to overlap.

The explanation for this is that such points approximate to the open-circuit voltage position where the current tends to zero irrespective of fixed or tracking process. Hence, the two graphs overlap.

4.2. Day 2

Table 2 summarises the results of the observations and measurements carried out on day 2. Unlike the previous investigation above, the measurements started 3 h earlier from 9.00 am.

Figure 4 shows the graphical comparison of the observed maximum power output for the fixed and the tracking systems.

It is important to emphasise that the pattern of the maximum power curves (**Figures 2** and **4**) does not take into account the behaviour of the power over the entire interval of the observations and measurements. It depicts the power under the I-V curves with maximum rectangular

Time	Vmp (V)		Imp (A)		Pmp (W)		Voc (V)		Isc (A)	
	Fixed	Trk	Fixed	Trk	Fixed	Trk	Fixed	Trk	Fixed	Trk
9 pm	3.0	3.0	0.6	0.72	1.8	2.16	0.03	0.03	0.63	0.75
10 pm	16.1	16.1	1.95	2.02	31.41	32.54	0.12	0.12	2.8	3.15
11 pm	15.0	15.0	2.54	2.67	38.1	40.05	0.13	0.13	3.27	3.56
12 pm	14.0	16.0	3.86	3.39	54.04	54.24	0.13	0.01	4.09	4.19
1 pm	14.0	16.5	2.86	2.45	40.04	40.43	0.01	0.15	3.57	3.79
2 pm	14.0	14.0	3.58	3.76	50.12	52.64	0.15	0.15	3.95	4.56
3 pm	10.5	10.5	2.87	3.5	3.14	36.75	0.15	0.15	3.0	3.84
Average maximum power (W)					31.24	36.97				
Percentage power gain by tracking (%)						18.34				

Table 2. Measurements and observations for day 2.

http://dx.doi.org/10.5772/intechopen.79710

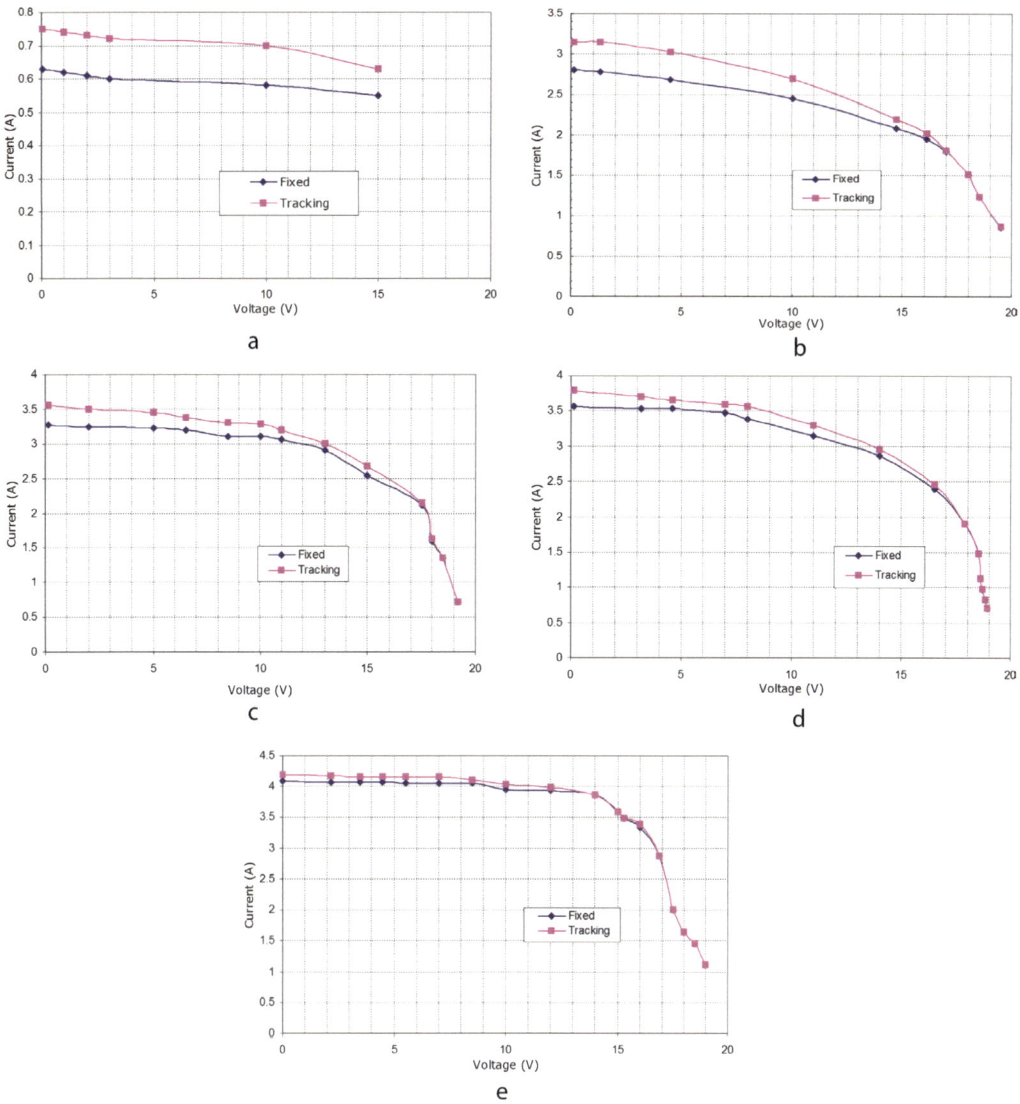

Figure 5. (a) I-V curves at 9.00 am (hor rad. - 223.7 KWh/m^2. Ambient temp: 14.6°C), (b) I-V curves for 10.00 am (hor rad. – 340 KWh/m^2. Ambient temp: 16.5°C), (c) I-V curves for 11.00 am (hor rad. - 499.9 KWh/m^2. Ambient temp: 19.23°C), (d) I-V curves for 12.00 pm (hor rad. – 576.2 KWh/m^2. Ambient temp: 20.48°C), (e) I-V curves for 1.00 pm (hor rad. – 606.4 KWh/m^2. Ambient temp: 21.2°C), (f) I-V curves FOR 2.00 pm (hor rad. – 653.1 KWh/m^2. Ambient temp: 21.67°C), (g) I-V curves for 3.00 pm (hor rad. – 629.5 KWh/m^2. Ambient temp: 22°C).

areas. For this reason, the power margin between the fixed and the tracking system at every point along the curves may appear closer as can be seen from the diagram above (**Figure 5**).

The closeness becomes more pronounced in a day with an even high insolation distribution. Comparison of the daily solar radiation for the first and the second day confirm that from the respective diagrams.

Observe the position of the curves in **Figure 5** above at 3.00 pm. At such position, the fixed system was beginning to lose site of sufficient direct radiation as the incident solar angle.

On the other hand, the tracking system was still busy following the Sun at that point as was observed in the field.

The I-V characteristics for day 2 were also plotted and compared for the two systems, fixed and tracking PV. **Figure 5** below show the characteristic curves.

The explanation for the above diagrams is the same with that of **Figure 4(a–e)**. The only notable difference is the appearance of the curves in **Figure 5a** at 9.00 pm.

The reason for this is that at that time for that particular day, the position of the Sun was such that the incidence angle was close to maximum and because the system output has been found to be inversely proportional to the incidence angle, hence the current for both the fixed and the tracking systems at that time appeared parallel to the voltage axis. The values of the current for both systems at the time were approximately the short-circuit currents of 0.63 A and 0.75 A for the fixed and tracking systems, respectively.

Recall that at the short-circuit current, the voltage becomes zero Eq. (4).

Hence, from the relation

$$P = IV \tag{6}$$

The power output for the systems approximated to zero. However, the maximum powers for the systems at that point as can be seen from **Table 2** are 1.8 and 2.16 W, respectively.

5. Conclusions

Key features of the research outcomes which contribute both to the aims of the study and knowledge are outlined below:

1. **Experimental Significance:** The study is absolutely an experimental work which involved a complete PV installation process for the fixed PV alongside the existing tracking PV originally used to power a water fountain. Based on the results of the I-V characteristics for the two systems: fixed and solar tracking PV in UK climate, it can be concluded that in UK and other locations with similar climatic conditions, both the fixed and tracking PV systems have a relatively slow response to sunrise. At noon time in UK, the performance of fixed PV systems approximates that of tracking PV systems. Also in the UK, fixed PV systems compared to tracking PV usually begin to lose sight of sufficient direct radiation after 3.00 pm, while the tracking system relatively remains further active as it still follows the Sun at such points

2. **Decision-Making:** The information gathered from this study can be used to reach decisions on the choice of either of the systems based on the electrical performance of both systems under the same insolation level and ambient temperatures. A common idea prior to this experimental study is that tracking PV generally out-matches the fixed systems. This no doubt is true however, from the results of the I-V characteristics, the margin between the electrical responses of both systems under similar conditions in the UK remain negligible for a longer part of the day. This implies that when the cost and maintenance for

http://dx.doi.org/10.5772/intechopen.79710

tracking PV systems are put into consideration at such location and other locations with similar climatic conditions, it may make more economic sense to choose the fixed option rather than the tracking PV option.

Finally, it has been shown that the materials used for solar cells in every solar PV module primarily determine the intrinsic efficiency of every solar PV module and system; Gallium arsenide was identified in the study to produce more efficient photovoltaic systems but much more expensive compared to silicon while polymetric or organic materials are much more cheaper but produce less efficient photovoltaic systems. It becomes inevitably necessary therefore to pay more attention to the research and development of cheaper and high efficient solar cell materials.

Author details

Chukwuemeka Ikedi

Address all correspondence to: ikedienergy@yahoo.com

Energy Commission of Nigeria, SunLab Technologies, Nottingham, UK

References

[1] Spanggaard H, Krebs FC. A brief history of the development of organic and polymeric photovoltaics. Solar Energy. 2004;**125**:146-183

[2] Hankins M. Solar Electric Africa. 2nd ed. Common Wealth Science Council; 1995

[3] Halliday D, Resnick R. Fundamentals of Physics. John Wiley and Sons Inc; 2013

[4] Luther J, Nast M, Fisch N, Christoffers D, Pfisterer F, Meissner D, Nitsch J. Solar Technology2008. DOI: 10.1002/14356007.a24_369

[5] Shivakumar R, Tippabhotla SK, Handara VA, Illya G, Tay AO, Novoa F, Dauskardt RH, Budiman AS. Fracture mechanics and testing of interface adhesion strength in multilayered structures. Solar Energy. 2016;**47**:55-89

[6] Dauskardt RH, Novoa D, Fernando C, David M. Environmental mechanisms of debonding in photovoltaic back sheets. Solar Energy. 2014;**87**:93-120

[7] Budiman AS, Illya G, Handara V, Caldwell A, Bonelli C, Kunz M, Tamura N. Enabling thin silicon technologies for next generation c-Si solar PV. Renewable Energy. 2014; **130**:303-308

[8] Bruno R, Bevilacqua P, Longo L, Arcuri N. Small size single-axis PV trackers: Control strategies and system layout for energy optimization. Solar Energy. 2015;**82**:737-743

[9] Bazilian M, Onyeji I, Liebreich M, MacGill I, Chase J, Shah J, Gielen D, Arent D, Landfear D. Re-considering the economics of photovoltaic power. Renewable Energy. 2013;**53**:329-338

[10] Huang BJ, Ding WL, Huang WC. Long-term field test of solar PV power generation using one-axis 3-position sun tracker. Solar Energy. 2011;**85**:1935-1944

[11] Koussa A, Cheknane A, Hadji S, Haddadi M, Noureddine S. Measured and modelled improvement in solar energy yield from flat plate photovoltaic systems utilizing different tracking systems and under a range of environmental conditions. Applied Energy. 2011;**88**:1756-1771

[12] James AR, Kauer M, Mellor AV, Paul N, Nicholas J, Ekins D. Current voltage characteristics of a metallic structure for a hot-carrier photovoltaic cell. PVSC. 2016. DOI: 10.1109/PVSC.2016.7749770

[13] Zhang Y, Gao S, Gu T. Prediction of I-V characteristics for a PV panel by combining single diode model and explicit analytical model. Solar Energy. 2017;**144**:349-355

[14] Ibrahim A. Analysis of electrical characteristics of photovoltaic single crystal silicon solar cells at outdoor measurements. Smart Grid and Renewable Energy. 2011;**2**:4965-4972

[15] Lorenzo E. Solar Electricity: Engineering of Photovoltaics. John Wiley and sons; 2000

[16] Ikedi CU. Numerical assessment of energy contribution by building integrated Photovoltaics in a commercial/office building refurbishment in UK. International Journal for Low Carbon Technologies. 2014;**11**:338-348